Her Story

A Timeline of the Women Who Changed America

CHARLOTTE S. WAISMAN, Ph.D.

& JILL S. TIETJEN, P.E.

Collins

An Imprint of HarperCollinsPublishers

HarperCollins books may be purchased for educational, business, or sales promotional use. For information, please write: Special Markets Department, HarperCollins Publishers, 10 East 53rd Street, New York, NY 10022.

FIRST EDITION

Designed by Laura Klynstra

Library of Congress Cataloging-in-Publication Data

Waisman, Charlotte S.
  Her Story : a timeline of the women who changed America / Charlotte S. Waisman & Jill S. Tietjen, P.E. Collins.—1st ed.
    p. cm.
  ISBN 978-0-06-124651-7
  1. Women—United States–History. 2. United States—History—Chronology. I. Tietjen, Jill S. II. Title.

  HQ1410.W354 2008
  973.082—dc22

                                        2007029942

Printed in China

08 09 10 11 12 ❖/ 10 9 8 7 6 5 4 3 2 1

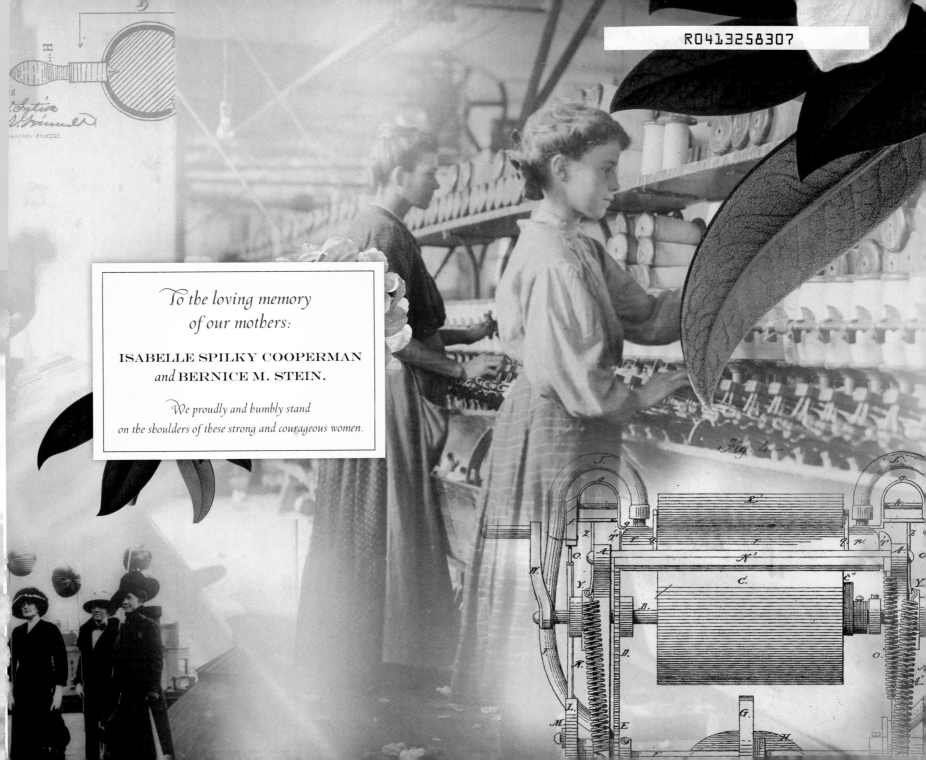

To the loving memory
of our mothers:

ISABELLE SPILKY COOPERMAN
and BERNICE M. STEIN.

We proudly and humbly stand
on the shoulders of these strong and courageous women.

Fig. 4.

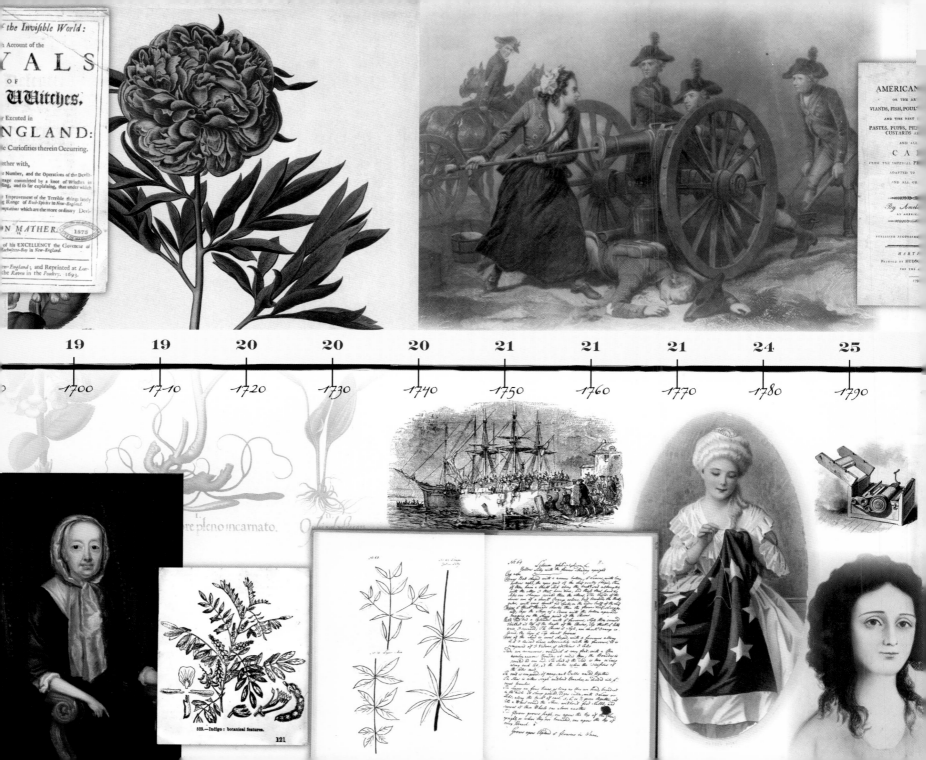

the Invisible World:
An Account of the
YALS
OF
Witches,
ly Executed in
ENGLAND:
le Curiosities therein Occurring.
ether with,
e Number, and the Operations of the Devils.
age committed by a knot of Witches in
ing, and so far explaining, that under which
e Improvement of the Terrible things lately
g Range of Evil-Spirits in New-England.
mptations which are the more ordinary Devil-
N MATHER.
1875
of his EXCELLENCY the Governor of
Machusetts-Bay in New-England.
ew-England; and Reprinted at Lon-
che Raven in the Poultry. 1693.

AMERICAN
OR THE AR
VIANDS, FISH, POULT
AND THE BEST
PASTES, PUFFS, PIE
CUSTARDS AN
AND ALL
CA
FROM THE IMPERIAL PR
ADAPTED TO
AND ALL GR
By Amel
AMER
PUBLISHED ACCORDING
HARTF
PRINTED BY HUDSO
FOR THE C

19	19	20	20	20	21	21	21	24	25

1700   1710   1720   1730   1740   1750   1760   1770   1780   1790

re pleno incarnato.

523.—Indigo: botanical features.
121

Table of Contents

PAGE No

16   16   16   17   17   17   17   18   18   18   19

1587   1600   1610   1620   1630   1640   1650   1660   1670   1680   169

Salem

Ye Ancient Plot of ye Towne of s'Gravesende 1645

0   93   106   118   131   145   164   187   208   221   229

10   1920   1930   1940   1950   1960   1970   1980   1990   2000   2007

27 28 29 34 39 43 50 56 64 72 8

1810 1820 1830 1840 1850 1860 1870 1880 1890 1900 1

Foreword by
MADELEINE ALBRIGHT

As a woman who has a long-standing, deep interest in history, political science, and foreign policy and as a former secretary of state (the first woman in U.S. history to hold this position), I am intrigued by this book by Charlotte Waisman and Jill Tietjen. It presents, in a unique time-line format, a history of important and influential American women—often women whose achievement was the first of its kind. I note as well, with satisfaction and appreciation, that many contemporary women of my acquaintance are included.

But it is the earlier women who continue to excite and interest me as I peruse this book. Until about thirty years ago, it was the accepted norm that history was the story of men. In fact, the history of the United States, from the arrival of the first colonists through the middle of the twentieth century, has been written almost exclusively by men. Early in U.S. history, events were often chronicled by religious leaders who described the times and recorded the contributions made by those in their particular group or by men in general. This means that, for the most part, the accomplishments of women were ignored, minimized, brushed aside, or even forgotten. Charlotte Waisman and Jill Tietjen's research brings the accomplishments of many women out of the shadows and onto the recorded pages of history.

I recall that the novelist Jane Austen has the heroine of her book *Persuasion*, Anne Elliot, argue that "men have had every advantage of us in telling their own story." The book you hold, *Her Story: A Timeline of the Women Who Changed America*, serves to make the tapestry of U.S. history richer by weaving women into that history. Women's significant accomplishments in every area of endeavor are described in the pages herein. It is important to recognize and celebrate these women, more than nine hundred in all, who are mostly unknown by the general population. No longer will the accomplishments of women throughout U.S. history be forgotten or remain invisible! Women have made significant and surprising contributions to our American way of life, and this book recognizes and celebrates them.

As cultures around the world educate and grant rights to women, the level of freedom and economic prosperity for all also rises, as can be seen by examples in our country. When women are recognized as leaders and welcomed into a given environment, that environment becomes better for all people. Furthermore, it helps us here in the United States to remember specifically the history of women's suffrage: after the signing of the Declaration of Independence, it took *more than 130 years* for women in the United States to gain the right to vote, with the ratification of the Nineteenth Amendment to the Constitution. That right was ensured in the end by a margin of only two men's votes (the mother of one insisted that he vote yes).

This right to vote was thus "given" to American women quite grudgingly and as late as 1920. I note in my most recent book, *The Mighty and the Almighty*, that one day the commander in chief, who sets the policy and tone of this country and speaks to the world on our behalf, will be a woman. And she will have reached that pinnacle of power due to the pioneering efforts of the women profiled in this book.

Our knowledge and recognition of the accomplishments of women in this country are so incredibly limited. What, for example, do the following items have in common?

- The computer compiler
- Paper bags at the grocery store
- Cultivation of indigo plants
- The Underground Railroad
- The Christian Science Movement
- A leukemia chemotherapy drug
- The American Red Cross

A woman invented, founded, or championed each! And most Americans today have never heard of these women. Accomplished women in their own right, Charlotte and Jill have turned their decades-long passion for women's history into a remarkable visual history lesson.

Spanning the centuries from 1587, when Virginia Dare was born in North Carolina, to the more familiar women of the twenty-first century, this book will allow women *and* men to become more aware of and informed about the women who have been instrumental in giving us the quality of life we enjoy today. Often stepping outside of the expected modes of behavior for women during their lives, the profiled women were the pioneers for their causes, their professions, or their passions. Their accomplishments have advanced the arts, the sciences, politics, and business.

Many of you are aware that while I was born in Czechoslovakia, I had the distinct privilege of immigrating to this country before World War II, while Europe was in turmoil. My family settled in Denver, Colorado, and I have many fond memories of the western "can-do" spirit. Women such as Charlotte and Jill epitomize that spirit and with this book demonstrate not only how women's roles in society have changed our history but also how as the women's roles have changed, our country has changed.

I noted earlier that as women around the world become educated, their right to vote is recognized and they become contributors to their country's economic prosperity. These changes will continue worldwide for women of the twenty-first century. In the United States, these changes will become increasingly more evident as women grow as *leaders*. Charlotte and Jill are actively helping women develop and become the leaders of tomorrow.

Charlotte, with several other women in Colorado, was instrumental in the creation of the Leadership Institute, which has now become the premier program of the Women's Vision Foundation. The Women's Vision Foundation, a truly unique Colorado organization, has for the past ten years been actively enhancing the success of corporate America by striving to make the corporate environment better for all employees. The vision of the foundation—women and men leading corporations that succeed because of their inclusive cultures—suits most admirably the basis of this book, which is a review of the myriad contributions by women in our society.

While Charlotte has been working with corporate women, Jill has been actively working to ensure that more women enter the professions as scientists, engineers, and technologists. She works with girls to expand their horizons and to make their economic self-sufficiency and career success an inevitability. To ensure that technical and pioneering women receive recognition for their efforts, she nominates them for such awards as the National Medal of Technology and the National Medal of Science and for induction into the National Women's Hall of Fame and the National Inventors Hall of Fame.

I am proud to know these two Colorado women and to support their work.

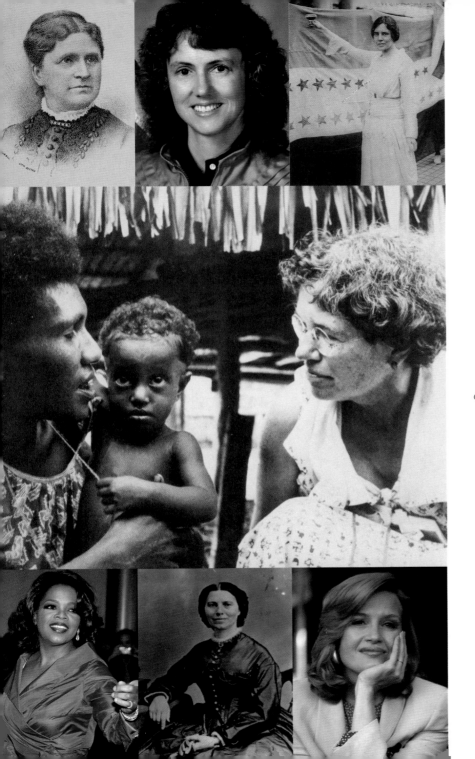

"Never doubt that a small group of thoughtful, committed citizens can change the world. Indeed, it is the only thing that ever has."

Anthropologist MARGARET MEAD

These words of Margaret Mead encapsulate the thought that the world is changed when someone gets an idea, becomes its champion, mobilizes others to support the idea, and then sees the idea through to implementation. Our book, *Her Story: A Timeline of the Women Who Changed America*, is the product of a small, thoughtful group of committed women who believed in us and wanted us to succeed in preserving a record of historical and current American women of achievement and note.

Our book is an educational and fascinating look at the far-reaching endeavors of more than nine hundred incredible women. It provides an understanding of women and their achievements throughout U.S. history; the context is a chronological timeline. This illuminating book format starts in 1587, when Virginia Dare was born in what is now North Carolina, and continues throughout U.S. history until the current day, covering many of the more well-known women of the twentieth and twenty-first centuries. The figures included here, whose accomplishments demonstrate the enormous range of achievements of

American women, were selected from a pool of more than three thousand candidates and include artists and entertainers, athletes, doctors and scientists, activists and politicians, pioneers and adventurers, corporate and business women, writers and journalists, educators, and entrepreneurs.

Included within the timeline are some snippets of information about societal and political events occurring at that same time, to provide historical context. The information in each entry is brief, providing a snapshot of an individual event or woman. This perspective allows the reader to move quickly from woman to woman and from achievement to achievement.

Although there are many books on women's history and some books that identify themselves as using a timeline format, none provides the intriguingly vivid, visual, and informational experience that *Her Story* does. Our goal is to pique your interest so that you will want to find out more about the significant women shown in our book and their many incredible accomplishments.

This book can be read easily, and the many pictures capture readers' interest and spur their curiosity. Some of the pictures are of the women described here, some illustrate the product or item they created, and some relate to the events that are noted. The images are the focus and heart of the timeline. It is our hope that the heavily illustrated format and the accessibility of the text will encourage both women and men to share this book with their children, especially their daughters, and encourage those daughters on the path to their own notable accomplishments.

We further believe our book will serve to record and preserve the legacy of the many remarkable women who came before us in the United States and on whose shoulders we in the twenty-first century stand. We hope you, our readers, will be inspired by the women profiled in this book and, in turn, will want to write both yourselves and your families into history.

Have you ever heard the phrase "Remember the Ladies"? In letters between Abigail Adams and her husband, then-congressman John Adams, in March 1776, the future First Lady encouraged her husband, a future U.S. president, to remember women while the Founding Fathers and their colleagues were putting together the governance structure of this country. She wrote, "I desire you would Remember the Ladies, and be more generous and favourable to them than your ancestors. Do not put such unlimited power into the hands of the Husbands. Remember all Men would be tyrants if they could." John's answer was that he could not help laughing at her "saucy" letter.

What he did not recognize was that with these words, Abigail became one of the first American women to clearly assert her desire for women's rights. The words of Abigail Adams have come to echo throughout U.S. history. For many other activists, including those in the suffrage movement, her words acted as a rallying cry for the equality of the sexes.

"Remember the Ladies" is still relevant to our times. By reading this book, you are honoring and remembering the women who came before us. We all have achieved so much because of the women profiled within its pages. It is no surprise that the year 2008 is much different for women (and men) than earlier times. The differences have been driven in significant part by the more than nine hundred women whose accomplishments are documented in this book. Many of the women and their achievements, particularly those in earlier years, have been forgotten or are invisible to the public at large.

We enjoy the rights and freedoms that we do today because of these earlier women and their courageous actions. At particular moments in America's history, some women chose to speak out in public for causes they believed in (education and women's suffrage, for example) despite the fact that in many situations women were not *allowed* to speak. Women had to fight for an education and for the right to own their own property—indeed, even to inherit property from their husbands or fathers. They fought to vote and to control their own reproduction. They often faced public ridicule, sometimes alone and sometimes with other women. We salute their courage and persistence.

In the twenty-first century it has become more common for women who are in their late teens and early twenties to believe that they are totally

MARGARET H. SANGER

equal to men and can do anything they wish. This news is exhilarating! Today's women can do it all *because* of what all of the women who came before them did to pave the way. Today's young women need to know who all of these brave, courageous, fearless, resilient, high-spirited women were, and we intend to tell them. The work that we have done in putting this book together has given us an awareness of the many awesome accomplishments of our foremothers.

A question we are frequently asked is "How did you select the women for inclusion in the book?" We chuckle somewhat in response. We began by making a long list of criteria and measured each woman and her achievement against it, but many times we broke our own "rules." In the final analysis, it was the two of us who considered, discussed, evaluated, and eventually chose. Our choices reflect our biases, as the works of earlier authors reflect theirs.

We fully believe that it is impossible to weigh one woman and her accomplishments against any other. Some of our choices were dependent upon the category of the contribution, its significance, its role as first of its kind, the time when the accomplishment occurred, or its overall contribution to American life. Sometimes we chose a woman because her influence and values touched a great number of people; sometimes we picked her because of the reverberations of her accomplishment. All these factors played a part in the selection process used in the making of this book.

Ultimately we came to understand that it is the very diversity of achievement that is critically important to identifying the breadth and depth of women's contributions to U.S. history. What we saw again and again is that the women portrayed in our book were often those who stepped outside of the expected modes of behavior for women during their lives. The profiled women were the pioneers for their causes, their professions, or their passions. Their accomplishments have both changed and advanced the arts, the sciences, politics, and business. These women are role models for all of us, and by featuring them in our book, we hope their successes will inspire other women to succeed as well. By representing women across the spectrum, we believe that any reader will be able to identify "women like me."

"How did you get the idea for this book?" We answer this frequently asked question most sincerely by saying that it's all due to a tea party and an essay contest. When we think back to these otherwise routine activities in our lives, we now understand that they were the first steps in our journey of writing this book and getting it published.

Charlotte attended a fancy tea party at the home of a friend. The guests played a parlor game in which they matched ten women's accomplishments with the names of the women. Charlotte was surprised that she was the only one who got all ten correct, but more important, she was appalled that she was the only one of the women in this highly educated group who could identify Margaret Sanger.

Who is Margaret Sanger? In 2008, women take for granted many basic rights of which our foremothers could barely dream: the right to vote, the right to own property, and the right to control the reproductive capabilities of our bodies. Margaret Sanger was instrumental in helping women achieve the last of these.

The parlor game and its results were a pivotal moment for Charlotte. She recognized that if this group of educated women did not know of Sanger, then it was possible that this woman would be unknown to much of the general population. What Sanger did, and the fact that she did it in the early twentieth century, was truly amazing. Charlotte believed that she needed to tell other people about women of achievement. She began to collect the names of the women who "came before" and to put each woman's accomplishment on a timeline.

At the time she and Jill met, there were more than three hundred historical women on this timeline. Charlotte used the timeline as part of a leadership seminar she taught, and each time she brought it out, seminar participants were enthralled with what they learned about the women who came before us.

Jill's odyssey began with an essay contest on the theme "great women in engineering and science." When the idea was proposed in 1987, Jill, like most Americans, could name only one great historical woman in engineering and science: Marie Curie. Curie was actually born in Poland and later resided in France. In order to provide the prospective essay writers, all sixth graders, with a list of great U.S. women in science and engineering from whom to select for their essay, Jill and her Society of Women Engineers colleague Alexis Swoboda had to do much research. Thus began Jill's collection of the names and accomplishments of women in history.

Jill used this information, as well as additional research, to create a number of volumes in her *Setting the Record Straight* series. She broadened her knowledge to include not only women with technical backgrounds but also many women educators, politicians, and activists. And like Charlotte, she continued to gather the names and stories of historical women.

So while we didn't know each other at the time, we were working from different perspectives in essentially the same field. *Connections* and *networking* describe most succinctly the way we met, worked, and developed the book you are now reading. We were introduced through the Women's Vision Foundation. Our connection came about through a woman who knew both of us separately and recognized that we simply *must* meet (thank you, Jill Marce). Since that meeting in early 2003, we have come to understand how important our complementary skills and interests have been to the success of this book. Our efforts together have led to this publication, about which we are passionate and of which we are extremely proud.

Concurrently with working on creating the book you are reading, we also produced a special timeline display that we used and will continue to use for touring around the country and speaking about the women in *Her Story*. We want the opportunity to speak with as many people as possible about the achievements of American women during the more than four hundred years since our country was founded.

Our traveling timeline is both large and graphically intriguing. In fact, it is *huge*—over a hundred feet long and more than three feet high. People viewing the exhibit are transfixed; they move down the timeline slowly, without speaking. Every person who views the exhibit comments about how much history he or she did not learn in school and never knew anything about. Another comment we hear frequently is that people are amazed at how many things happened earlier than they expected—and, at the same time, that other things happened much later than they would have expected. The diversity of the women presented in the timeline and the range of their achievements are also mentioned by those who have seen our traveling exhibit.

We expect our book to be the start of many discussions and conversations related to women's achievements throughout history. This is a book for women and their daughters. It is a book for students in elementary school through college. It is a gift book: men to women, women to women, and men and women to their children. *Her Story* captures the broad spectrum of human endeavor and is a book that will allow women and men to become informed about and proud of the history that enables us to have the quality of life that we enjoy today.

We are indeed proud to offer you *Her Story: A Timeline of the Women Who Changed America*. After you read and savor it, take the time to write *your* family into history. In the appropriate year, honor and memorialize such wonderful accomplishments as the first woman in your family who finished her undergraduate degree, wrote a book, or raised five children as a single mother (and sent them all to college). Charlotte has seen the pride evidenced by women in her leadership classes who acknowledge the achievements of their predecessors by writing them into history through the timeline. The many stories we have heard are a superb testimony to the American spirit—thoughtful, committed women do indeed change the world!

1587

One hundred sixteen people, including Eleanor and Ananias Dare, settle on Roanoke Island (now part of North Carolina) as part of the second effort of the British to colonize the New World. Their daughter, *Virginia Dare*, is the first child to be born to English parents on what is now American soil. Unfortunately, this entire group vanished. A U.S. stamp commemorating Virginia Dare's birth and the "Lost Colony" is later issued.

1608

Pocahontas saves a Jamestown colonist, Captain John Smith, from execution by her father, Algonquin chief Powhatan.

Purchased by the Virginia colony, the first African slaves (seventeen men and three women) arrive at Jamestown. By the end of the transatlantic slave trade, an estimated five hundred thousand people have been "exported" to the United States; 30 percent of them are women.

1619

-1580   -1590   -1600   -16-10

1607

One hundred four British men and boys settle in Jamestown (later part of the state of Virginia) as part of continuing attempts by the British to colonize the New World. Jamestown is named in honor of James I, the king of England. This group of pioneers was sponsored by the Virginia Company of London.

1620

More than a hundred Pilgrims, men, women, and children, leave England aboard the ship *Mayflower* in search of religious freedom. When the ship arrives at Plymouth Rock, its passengers establish the first settlement in what will become the state of Massachusetts.

1622

Mary Johnson

A slave named Mary arrives in Jamestown; she later becomes one of the first freed slaves in the United States who has the right to choose her own surname (Johnson). A common practice at the time is to choose the surname of a benefactor; thus it is believed that someone with the surname of Johnson helped Mary and her husband attain their freedom.

1644

Upon the death of her husband, *Mistress Sarah Jenney* becomes the first woman to run a grain mill. In 1652, this Pilgrim and resident of the Plymouth Colony receives a land grant.

Margaret Brent emigrates from England to the Maryland settlement in 1638 with one of her sisters and two of her brothers. She claims a land grant, conducts business, and appears in court. On his deathbed, Governor Leonard Calvert appoints her as his executor, and her ensuing actions enable the settlement to survive. In 1648, she appears before the legislature to request two votes—one for herself, as she is a landowner, and another in her role as Lord Baltimore's attorney.

1648

1620   1630   1640   1650

1636

Harvard College is established in Massachusetts; its mission is to educate men to become ministers.

1650

Anne Bradstreet, who emigrates from England to Massachusetts with the Puritans in 1630, is the first American poet and the first American woman to have a book published in what will become the United States. She writes poetry while raising eight children and serving as hostess for her husband, who becomes governor. Her first book of poetry, *The Tenth Muse Lately Sprung up in America, by a Gentlewoman of Those Parts*, is published in England in 1650 by her brother-in-law, who takes her poems without her knowledge. The American edition is published in 1678.

1637

Anne Hutchinson expresses religious views that are not appreciated by some influential members of the local government. Their accusations result in her becoming the first female defendant in a Massachusetts court and, upon a finding of her guilt, her expulsion from the Massachusetts Bay Colony in 1638. She and much of her family settle in Rhode Island.

1645

Lady Deborah Moody founds Gravesend (now a neighborhood in Brooklyn, New York) based on a town patent she receives from the Dutch. Moody becomes the first female landowner in the New World, which also makes her eligible to vote.

Her Story

Margaret Hardenbroeck Philipse emigrates from Holland to New Amsterdam (now New York). She is a highly successful entrepreneur who ultimately comes to own most of the real estate in Westchester County. She also owns a number of ships, which she uses in the fur trade. This first woman business agent in the colonies raises five children and is the wealthiest woman in New Amsterdam.

1682

The account by *Mary Rowlandson* of her captivity by the Narragansett Indians is published and becomes an immediate best seller.

1660

1655

At a time when all women are viewed as the property of a man and most black women in the New World are enslaved, the slave *Elizabeth Key*, daughter of a free white man and an enslaved woman, sues for freedom. Hers is one of the earliest such suits filed in the English colonies. Through a series of owners, she is enslaved for a total of nineteen years, although a previous owner had committed to releasing her from bondage after nine years of service. Two courts rule against her; through petition, her case is sent to the Virginia General Assembly. Elizabeth is ultimately freed.

Kateri Tekakwitha is known as "Lily of the Mohawks"; she converts to Catholicism in 1676. The first Native American on the path to sainthood in the Roman Catholic Church, she is declared venerable in 1943 and beatified in 1980.

1676

1650   1660   1670

1660

Quaker *Mary Dyer* is imprisoned in 1657 in Massachusetts because of her religious beliefs and expelled from the New Haven colony in 1658 for preaching about them. When she is hanged in Massachusetts in 1660 for defying a law banning Quakers from the colony, she becomes the only woman in U.S. history to die for religious freedom. There is now a statue in her memory on the grounds of the State House in Boston.

Pennsylvania is founded as a Quaker haven. **1681**

The Wonders of the Invisible World:

Being an Account of the

TRYALS

OF

Several Witches,

Lately Executed in

NEW-ENGLAND

And of several remarkable Curiosities therein Occurring.

Together with,

I. Observations upon the Nature, the Number, and the Operations of the Devils.
II. A short Narrative of a late outrage committed by a knot of Witches in Swede-Land, very much resembling, and so far explaining, that under which New-England has laboured.
III. Some Councels directing a due Improvement of the Terrible things lately done by the unusual and amazing Range of Evil-Spirits in New-England.
IV. A brief Discourse upon those Temptations which are the more ordinary Devices of Satan.

By COTTON MATHER.

Published by the Special Command of his EXCELLENCY the Governour of the Province of the Massachusetts-Bay in New-England.

Printed first, at Boston in New-England; and Reprinted at London, for John Dunton, at the Raven in the Poultry. 1693.

1692

1692 More than a hundred people, mostly women, are accused of witchcraft; twenty men and women are executed.

17-10

1704 Author *Sarah Kemble Knight* keeps a journal of a difficult and long trip she makes from Boston to New York and back, a trip not usually taken by a woman of the day. After the diary's publication in 1825, it is acclaimed for its depiction of Knight's encounters with the people and places of the time.

Her Story

1712

Hannah Callowhill Penn, the wife of William Penn, manages the affairs of Pennsylvania and plays an important role in holding the province together during her husband's long illnesses and after his death. She serves as the acting proprietor of the Province of Pennsylvania from 1712 until her death in 1726. She is the first woman given the status of honorary citizen of the United States (1984).

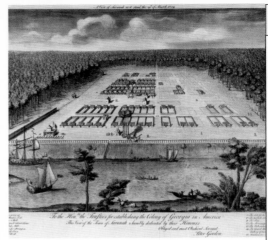

1733

Born of a white trader and a Creek Indian of royal blood, *Mary Musgrove* establishes with her husband a successful trading post near modern-day Savannah, Georgia. She serves as a negotiator and interpreter for General James Oglethorpe in 1733 when he brings the first English colonists to what will become the state of Georgia. Prior to her death, she is Georgia's largest landowner by a grant from the British Crown.

1710   *1720*   *1730*

1744

1715

Sybilla Masters, an early woman inventor, patents a method of crushing corn at the gristmill she and her husband own. Masters is believed to have developed the process after watching Tuscarora Indian women beat corn with a pestle in large bowls. She travels to London to secure the patent. As with all patents of the time, it is issued to a man (her husband).

The technique for large-scale cultivation and processing of indigo for dye that *Eliza Lucas Pinckney* developed as a teenager, when she was already manager of her father's extensive plantation holdings, achieves its first commercial success. Indigo sales support the Carolina economy for the next thirty years. Later, this outstanding businesswoman, wife, and mother is recognized as a leading member of Carolina society. Both her sons become leaders in the newly independent United States: one signs the U.S. Constitution and the other is governor of South Carolina. President Washington serves as one of her pallbearers. Pinckney writes: "I have the business of three plantations to transact, which requires much writing and more business and fatigue of other sorts than you can imagine. But least you should imagine it too burdensome to a girl at my early time of life. . . I assure you I think myself happy."

Poet *Lucy Terry Prince* writes about an Indian raid in Deerfield, Massachusetts. "Bars Fight" is the earliest existing published poem by a former slave.

1746

Martha Daniell Logan, a horticulturist, begins writing a column, "Gardener's Kalendar," in which she discusses herbs, vegetables, flowers, gardens, and orchards. Her column continues to be published for forty years.

1751

Cherokee **Nancy Ward** is given the title of distinction "Beloved Woman" (Ghighua) for great bravery in battle and leading her people to victory. Accompanying the title are responsibilities such as a seat at the General Council of Chiefs, leadership of the Women's Council, and a role as ambassador. She counsels the Cherokee tribe against land cession and speaks on behalf of her people with U.S. representatives. No other Cherokee woman will be honored with this title for more than a hundred years.

1755

1758

Fourteen-year-old **Mary Jemison** is adopted by Seneca Indians after being captured in a raid at her family farm during which most of her family is killed. Unlike most other settler women who are captured by Indians, Jemison stays with the Senecas and learns to value and honor their customs. She becomes well known for her acceptance of all people and rises to high social standing in the tribe. She is issued a tribal grant and owns one of the largest herds of cattle in the region. After her death at age ninety, a statue is erected in her memory.

1750

1760

1770

1757

Jane Colden, the first American woman botanist, uses ink impressions of leaves and sketches living plants to catalogue more than three hundred species. She also discovers and names the gardenia, a flowering plant.

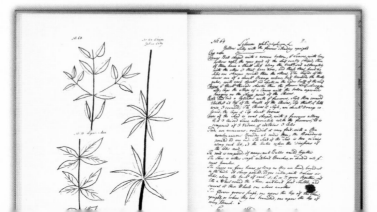

1769

Patience Wright is America's first professional sculptor. Newly widowed and with five children, she begins her career as a wax modeler, displaying much skill. She puts together a traveling waxwork exhibit of contemporary famous people–a new type of exhibit (predating Madame Tussaud by thirty years). After her New York exhibit is destroyed in a fire in 1771, she moves to England, where she creates other sculptural art.

Her Story

1773 Patriot *Sarah Fulton* is an active freedom fighter. During the Boston Tea Party, when Boston patriots dress up as Mohawk Indians to protest the imposition of a tax on tea, Fulton is there to help with the costumes and makeup removal. After the Battle of Bunker Hill, Fulton helps nurse the wounded soldiers. She delivers a dispatch from General Washington to the Revolutionary troops by walking to the Charles River and then rowing across. She is later honored by visits from both General Washington and the Marquis de Lafayette. After her death, Fulton Street in Medford, Massachusetts, is named in her memory.

British troops landing in Boston lead to battles in Lexington and Concord, Massachusetts–the first shots of the American Revolution.

1775

1773 One-third of the slaves brought into the United States are women. Emancipated slave and poet *Phillis Wheatley* writes of their suffering in *Poems on Various Subjects, Religious and Moral* and is the first African American woman to publish a book.

1774

POEMS

ON

VARIOUS SUBJECTS,

RELIGIOUS AND MORAL,

BY

PHILLIS WHEATLEY,

NEGRO SERVANT to Mr. JOHN WHEATLEY,
of BOSTON, in NEW ENGLAND.

LONDON:
Printed for A. BELL, Bookseller, Aldgate; and sold by
Messrs. COX and BERRY, King-Street, BOSTON.

MDCCLXXIII.

Published according to Act of Parliament, Sept.ʳ 1.1773 by Arch.ᵈ Bell.
Bookseller Nº 8 near the Saracens Head Aldgate.

PHILLIS WHEATLEY, NEGRO SERVANT to Mr. JOHN WHEATLEY, OF BOSTON.

1773

Poet, dramatist, and historian *Mercy Otis Warren* advocates for national independence from royal tyranny in her satirical play *The Adulateur*, published in a Boston newspaper. She subsequently corresponds with Abigail Adams (later First Lady), expressing her opinion that women do not participate in all matters of life not because of their lack of ability but primarily because of the lack of suitable education and opportunity for them to pursue their talents.

Appointed postmaster of Baltimore, Maryland, and probably the first woman in the colonies to hold such a position, printer and newspaper publisher *Mary Katherine Goddard* remains in this position until 1789. At that time, she is replaced against her will, and against the wishes of the populace, with a man who could do the traveling the position required, which by

the mores of the time women "could not manage." In 1777 she issues the first printed copy of the Declaration of Independence to include the names of all of the signers. Later she issues an almanac in her own name and also operates a bookstore.

After coming to New York State in 1774 from England, *Mother Ann Lee* founds the religious movement known as the Shakers, so called because of participants' singing, dancing, shouting, shaking, and speaking in tongues. Lee advocates for equal rights and responsibilities for women and men, an equalitarian order, and the dignity of labor.

Betsy Ross sews the first American flag.

1775

1776

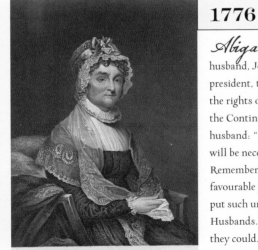

A. Adams

Abigail Adams encourages her husband, John Adams, who in 1797 becomes president, to "Remember the Ladies" by considering the rights of women during the deliberations of the Continental Congress. Adams writes to her husband: "In the new Code of Laws which I suppose will be necessary for you to make I desire you would Remember the Ladies, and be more generous and favourable to them than your ancestors. Do not put such unlimited power into the hands of the Husbands. Remember all Men would be tyrants if they could. If particular care and attention is not paid to the Ladies we are determined to foment a Rebellion, and will not hold ourselves bound by any Laws in which we have no voice, or Representation."

New Jersey legalizes a woman's right to vote. Women in all other states lose this right through the enactment of the U.S. Constitution (signed in 1787, ratified in 1789). In reaction to large numbers of women voting in the 1800 presidential election, New Jersey rescinds a woman's right to vote in 1807 by inserting the words *white* and *male* into the requirements for voters.

Her Story

1777

Sixteen-year-old *Sybil Ludington*, dubbed the "female Paul Revere," rides all night in heavy rain, eluding both highway robbers and the British. She knocks on farmhouse doors in Putnam County, Connecticut, for forty miles to warn of the British approach. Four hundred men arrive in time to fight and win the Battle of Ridgefield, driving the British back to Long Island Sound. General George Washington later congratulates her for her bravery.

"Mum Bett," an illiterate slave later known as Elizabeth Freeman, overhears conversations about freedom for men while she is serving food and beverages. Determined to seek her own freedom, she finds an attorney willing to take her case and the case of another slave of the Ashley family. As a result of *Brom & Bett v. Ashley*, which is argued before a Massachusetts county court, she becomes one of the first slaves to be freed under the Massachusetts constitution of 1780. Because of this decision, the freedom of the remaining Massachusetts slaves is also secured. A great-grandchild of Elizabeth Freeman, born many years later, will be the famous writer W. E. B. Du Bois.

1781

1777

1778

During the Battle of Monmouth, *Molly Pitcher* (Mary McCauley) fights in her husband's place after his death. She becomes a sergeant and receives a military pension for the rest of her life.

1780

1780

"Daughter of Liberty" *Esther DeBerdt Reed* organizes a women's committee in Philadelphia, Pennsylvania, to raise money for General Washington's troops. They raise the extraordinary amount of $300,000 in Continental (paper) dollars from more than sixteen hundred contributors by going door-to-door.

1785

1779

The Continental Congress grants a lifetime pension to Revolutionary War heroine *Margaret Corbin*, who was disabled in 1776 after taking up her fallen husband's gun during the defense of Fort Washington.

Playwright and actress *Susanna Haswell Rowson* publishes the first American best-selling novel, *Charlotte Temple: A Tale of Truth.* Her work is the first written specifically for women readers and is very popular at the time.

1790

1790

1782

Deborah Sampson is one of several women known to have fought in the Revolutionary War by impersonating a male. She later receives a pension for her valiant service. After her death, her husband is granted a survivor's pension.

DEBORAH SAMPSON.
Published by H. Mann. 1797.

1790

Judith Sargent Stevens Murray writes and publishes articles on equal rights for women; she is an early champion of female equality.

Her Story

PAGE Nº 25

School graduations offer one of the few opportunities young women at this time have to speak in public.

1794

Elizabeth Ann Bayley Seton is part of a group that founds the Society for the Relief of Poor Widows with Small Children; she serves as its treasurer until 1804. After converting to Catholicism in 1805, Seton founds the Sisters of Charity of St. Joseph, the first American sisterhood, in 1809 in Maryland. The funds from the boarders at St. Joseph's School enable her to offer free schooling to needy girls, thus leading to her being called the "foundress of the parochial school system in the United States." She is canonized as a Roman Catholic saint in 1975.

1807

Catherine Greene gives Eli Whitney the idea of a brush to sweep away cotton seeds; he gets full credit for the invention of the cotton gin.

1793

Catherine Ferguson opens New York City's first Sunday school. Born a slave, with her mother sold when she was eight, Ferguson is sensitive to the needs of destitute children. The Murray Street Sabbath School, integrated from its founding, serves poor children for forty years.

1797

1790

1795

1800

Amelia Simmons publishes the first American cookbook.

1796

1800 The United States has the highest recorded birthrate in the world at 7.04 babies per woman.

AMERICAN COOKERY,

OR THE ART OF DRESSING

VIANDS, FISH, POULTRY and VEGETABLES,

AND THE BEST MODES OF MAKING

PASTES, PUFFS, PIES, TARTS, PUDDINGS, CUSTARDS AND PRESERVES,

AND ALL KINDS OF

CAKES,

FROM THE IMPERIAL PLUMB TO PLAIN CAKE.

ADAPTED TO THIS COUNTRY,

AND ALL GRADES OF LIFE.

By Amelia Simmons,

AN AMERICAN ORPHAN.

PUBLISHED ACCORDING TO ACT OF CONGRESS.

HARTFORD:

PRINTED BY HUDSON & GOODWIN,

FOR THE AUTHOR.

1796.

1805

While accompanying her husband, *Sacagawea* serves as a guide and interpreter for the Lewis and Clark expedition, whose mission is to find a water route through North America and explore the uncharted West. During this journey of over two years, the Shoshone woman takes on an increasingly important role because since she and her baby accompany the group, hostile Indian tribes recognize that the expedition is peaceful. It is believed that more mountains, lakes, and streams bear her name than that of any other North American woman.

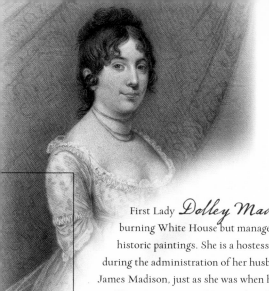

1814

First Lady *Dolley Madison* flees a burning White House but manages to save many historic paintings. She is a hostess extraordinaire during the administration of her husband, President James Madison, just as she was when he was Thomas Jefferson's secretary of state and as she would be for many years thereafter.

1819

After her husband's death, *Kaahumanu* becomes co-ruler of Hawaii with the new king, Kamehameha II. She is an active proponent of fair law and uses her power to abolish many restrictions on women. She encourages the new king to eat publicly with women, a previously forbidden act. In 1824, she proclaims Hawaii's first code of laws that prohibit murder, theft, and fighting. She orders that schools be created for all people to learn to read and write, and she establishes trial by jury.

Of 871,000 black women in the United States, 86 percent are slaves.

1820

-18-10

-18-15

-18-20

1813

Mary Young Pickersgill

makes the flag that serves as the inspiration for Francis Scott Key's "The Star Spangled Banner."

1816

Factory workers in the United States number one hundred thousand. Two-thirds of them are women and children, who are paid significantly less than men.

1809

Mary Kies receives the first patent awarded to a woman. The device she invented was for weaving straw with silk or thread.

Her Story

PAGE № 27

1821

Emma Willard opens the Troy Female Seminary in New York. Next to last of seventeen children, she was encouraged by her father to acquire an education beyond that generally expected for a girl of her time, especially a farm girl. Her progress was so rapid at school that by the age of fifteen, she was teaching others. She opened the Middlebury Female Seminary in Vermont in 1814, demonstrating that women could teach and girls learn classical and scientific subjects thought only appropriate for males to study. The Troy school ultimately becomes one of the most influential schools for women in the United States. In 1895, the school is renamed the Emma Willard School; it is still in operation today.

1820   1822

1824

Author *Margaret Bayard Smith* publishes her first book about Washington society, based on reminiscences and real incidents. Her letters and notebooks from the period 1800 to 1841, published in book form in 1906, cover U.S. history from the presidency of Jefferson to that of Jackson and provide significant insights into U.S. social and political history.

1825

Rebecca Webb Lukens assumes the helm of the Pennsylvania iron mill formerly run by her father and husband. Though the mill is in dire financial straits, she quickly reinvigorates and expands the business. She is the first woman in America to be active in the iron business and probably the first in any heavy industry. The Lukens Steel Company remains in operation today.

1829

Frances (Fanny) Wright, who emigrated to the United States from Scotland and is a strong advocate of women's rights, birth control, equal education for men and women, and the abolition of slavery, assists in the formation of the Association for the Protection of Industry and for the Promotion of National Education, which is an early labor union.

To please the U.S. government, the Cherokee Nation adopts a new constitution that eliminates women's power in decision making by denying them the right to vote.

1827

1824

1824

At a time when women artists are frowned upon, sisters and artists *Anna Claypoole Peale* (pictured) and *Sarah Miriam Peale* are both elected to the Pennsylvania Academy of the Fine Arts. Anna is a miniaturist. Sarah, primarily a painter of portraits in oil, achieves a successful career in the arts never before reached by a woman in America. Not only are they the first women to achieve full professional standing and recognition as artists, but they reach this position early and sustain it for more than fifty years.

1828

1830

1831

The first American-born female lecturer, *Maria Stewart* is recognized for delivering several public addresses in Boston at a time when very few women even speak in public. Stewart, like most African American women of the time, received no formal education except what she gleaned attending Sunday school. In her biblically influenced speeches, Stewart encourages blacks to become educated and to press for their rights. In a political essay addressed to the "daughters of Africa" and published in *The Liberator,* she says: "Possess the spirit of independence. . . . Sue for your rights and privileges. Know the reason that you cannot attain them. Weary them with your importunities. You can but die if you make the attempt; and we shall certainly die if you do not."

Her Story

1833

Abolitionist and suffragette *Martha Coffin Wright*, the sister of Lucretia Mott, attends the founding meeting of the American Anti-Slavery Society in Philadelphia. While six months pregnant and at a time when pregnant women did not appear in public, she helps plan the 1848 Seneca Falls women's rights convention. Her home in Auburn, New York, is part of the Underground Railroad and her neighbors call her "a very dangerous woman." In 1863, she helps found the Women's National Loyal League, whose abolitionist objectives are achieved with the ratification in 1865 of the Thirteenth Amendment, which abolished slavery.

1832

Maria Weston Chapman joins with twelve other women to form the Boston Female Anti-Slavery Society.

1833

Author, editor, and prominent abolitionist *Lydia Maria Child* publishes her first antislavery book, *An Appeal in Favor of That Class of Americans Called Africans*. The book contains a proposal to right social wrongs by educating all African Americans. At that time, Child is already a prominent writer: one of her books, *The Frugal Housewife*, goes through at least thirty-five printings between 1829 and 1850. Child's antislavery books cost her the support of many of her readers; she suffers social ostracism and her membership in the prestigious Boston Athenaeum is revoked.

1833

Prudence Crandall, a highly educated white woman, scandalizes society when she establishes an academy for black girls; in 1995, by an act of the state General Assembly, she becomes "Connecticut's state heroine." Prudence Crandall Hall, a dormitory at Howard University and part of the Harriet Tubman Quadrangle, is named in her memory.

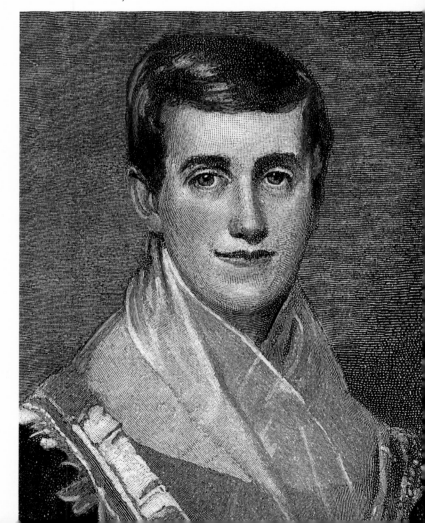

1833

Oberlin College admits its first class of forty-four students; fifteen of them are women. Women, however, are not allowed to study the same curriculum and earn an A.B. degree (like the men) until 1837.

1834

Textile workers in Lowell, Massachusetts—mostly young women—strike to protest their wages being cut by 15 percent. At this time in America, there are few occupations where women can earn their own money. Textile mills provide paid work outside the home for many young women, though the mills are not safe and the working conditions are deplorable. A strike, especially one that is organized and led by women, is a most astounding event at this time.

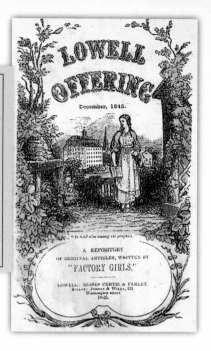

1834

1833

Maria Martin Bachman
paints the plants, flowers, insects, and other details for John James Audubon's masterpiece *Birds of America*.

1835

1835

1835

Enduring mob violence,
Paulina Kellogg Wright Davis
organizes an antislavery convention in Utica, New York.

Physician and reformer *Harriot Kezia Hunt* opens her own medical practice, concentrating in physiology and having primarily women and children as clients. Because she is female, she is barred from the hospitals. She tries twice to attend lectures at Harvard Medical School but is denied permission because of her gender. The national publicity resulting from Harvard's failure to admit her leads to the Female Medical College of Pennsylvania awarding her an honorary degree in 1853. Her focus continues to be on women's health issues as well as equal rights for women. For twenty years, her annual tax payments to Boston authorities are accompanied by a protest of taxation without representation because she cannot vote; the protests are published in the newspapers.

1837

Mount Holyoke Female Seminary (which will become Mount Holyoke College in 1888) opens, due to the efforts of *Mary Lyon*. Lyon began teaching in country schools in order to finance her own education. She recognized that there was no clear curriculum for women's education and that what they studied was too dependent on teachers who might be ill-prepared. Strongly determined, Lyon started raising funds for a female seminary with high academic standards. Students had to pass an exam to be admitted and were required to help with domestic chores. So great is the demand for this educational program that in 1838, more than four hundred applicants are turned away. After the school opens, Lyon acts as a principal and a teacher and organizes the domestic work system that pupils use to raise their own tuition. Lyon says: "It is one of the nicest of mental operations to distinguish between what is very difficult and what is utterly impossible."

1836

Narcissa Prentiss Whitman travels with her husband, a Presbyterian minister, from the East Coast to Oregon. She rides sidesaddle during this rugged trip and is one of the first two white women to cross the Continental Divide, at South Pass, Wyoming. In 1912, an opera, *Narcissa*, is based on the story of the difficulties of this journey.

1837

Phoebe Worrall Palmer is one of the founders of the Holiness movement in Christian fundamentalism.

1837

Sarah Josepha Hale, who initially turned to writing as a way to support her five children after the death of her husband, is selected to be the editor of a new monthly women's magazine, *Godey's Lady's Book,* the first issue of which appears in January. For forty years (until age ninety) she will fight for greater educational opportunities for women through its editorial columns.

1838

Rebecca Gratz founds the first American Jewish Sunday school and Jewish orphanage. She is also believed to be the model for the heroine Rebecca in Sir Walter Scott's novel *Ivanhoe.*

1837

1838

1838

Abolitionist and women's rights pioneer *Angelina Grimké,* who left a life of southern affluence to become a Quaker and live in Philadelphia, where she became an active member of the abolition movement and lectured and wrote on the evils of slavery, becomes the first woman to address a legislative body when she asks the Massachusetts legislature to end the slave trade in the state.

1838

Abolitionist, suffragist, and reformer *Abby Kelley Foster,* who grew up as a Quaker, developed a spirit of independence and a high moral commitment, and became an effective public speaker against slavery, embarks on a speaking tour. She is denounced from the pulpit in one community, forbidden to speak at another stop, and described as a "bad woman" by a hotelkeeper at a third. She continues to commit herself totally to reform despite these and other criticisms and to speak out for the cause.

Her Story

The school initially chartered as the Georgia Female College is one of the oldest institutions of higher education for women in the United States and the first college in the world chartered to grant degrees for women; its first graduates receive bachelor's degrees in 1840. It is later renamed Wesleyan College and remains an all-female college to this day.

1840

1839

1839

Margaret Fuller starts "Conversations" for women, a salon to discuss issues and ideas. Fuller is a brilliant and highly educated woman who is influenced by Ralph Waldo Emerson and Transcendentalism. She becomes editor of *The Dial*, the Transcendentalist magazine and contributes a number of essays, reviews, and poems. In 1845, she publishes her pioneering and classic feminist work *Woman in the Nineteenth Century*, which contributes significantly to the women's rights movement.

1839

Abolitionist and former South Carolinian *Sarah Grimké*, with her sister Angelina and Angelina's husband, publishes *American Slavery As It Is: Testimony of a Thousand Witnesses*. The volume serves as a sourcebook for Harriet Beecher Stowe's *Uncle Tom's Cabin*.

1840

1840

Businesswoman and philanthropist *Margaret Haughery* opens the New Orleans Female Orphan Asylum, which she maintains with proceeds from her dairy. She later sells the dairy and acquires a bakery, establishing the first steam bakery in the South and developing the innovation of packaged crackers. Her bakery becomes the city's largest export business. The "Bread Woman of New Orleans" is so beloved that in 1884 a statue is dedicated to her. Each year on February 9, New Orleans celebrates Margaret Haughery Day.

1840

Influential Quaker, abolitionist, and women's rights pioneer *Lucretia Mott* meets Elizabeth Cady Stanton in London, at the World Anti-Slavery Convention, where all women are denied seats. This bonding experience leads the women to organize the famous Seneca Falls (New York) women's rights convention in 1848. Later, Mott is named president at the first convention of the American Equal Rights Association (a women's suffrage organization) in 1866. For more than fifty years, she is one of the most consistently effective women to play a major role in social reforms. Mott said: "The world has never yet seen a truly great and virtuous nation because in the degradation of woman the very fountains of life are poisoned at their source."

1840

Reformer and feminist *Ernestine Rose* begins work on legislation granting married women property rights. For over two decades she campaigns tirelessly for women's rights. Between 1850 and 1870, she lectures in more than twenty states, addresses legislative bodies, and attends most national and state women's rights conventions. Her lectures address the issues of antislavery, temperance, and freedom of thought with women's rights.

1843

1843

Preacher Isabella Baumfree, who gains her freedom in 1827, changes her name to *Sojourner Truth.* She becomes a powerful antislavery speaker best remembered for her 1851 "Ain't I a Woman" speech in which she said: "That man over there says that women need to be helped into carriages, and lifted over ditches, and to have the best place everywhere. Nobody ever helps me into carriages, or over mud-puddles, or gives me any best place! And ain't I a woman? Look at me! Look at my arm! I have ploughed and planted, and gathered into barns, and no man could head me! And ain't I a woman? I could work as much and eat as much as a man–when I could get it–and bear the lash as well! And ain't I a woman? I have borne thirteen children, and seen most all sold off to slavery, and when I cried out with my mother's grief, none but Jesus heard me! And ain't I a woman?"

1843

Humanitarian *Dorothea Dix* exposes the harsh treatment of the mentally ill when her report documenting an eighteen-month survey of facilities throughout the state is presented to the Massachusetts legislature. The report speaks of horrible conditions including filth, cruelty, and disease. Dix presents reports in many other states, and her careful research and public speaking lead to the construction of the New Jersey State Lunatic Asylum in Trenton in 1848, which she dubs her "firstborn child." The institution becomes a model for the humane care of the mentally ill. In 1843, there are 13 mental hospitals in the United States; by 1880, due to Dix's efforts, there are 123. Her efforts also help provide the foundation for enhancements in diagnosis and treatment of mental illnesses.

I Sell the Shadow to Support the Substance.

SOJOURNER TRUTH.

1844

Businesswoman, humanitarian, and landowner *Juana Briones* is a Mexican American pioneer in the San Francisco, California, area who prospers and raises eight children on her own. She is a healer, a nurse, and a midwife, all without formal training. She uses herbal medicine to care for her family and others. She has a garden and cows, and sells milk and vegetables to ship's crews. She purchases 4,400 acres to expand her cattle and farming interests. Her fame as a healer and her generosity make her a role model and a legend.

1844

Fanny Crosby, who was blinded at the age of six weeks, has her first book of poems, *The Blind Girl and Other Poems,* published. The author of the well-known hymn "Blessed Assurance" will write more than 5,500 hymns in her lifetime.

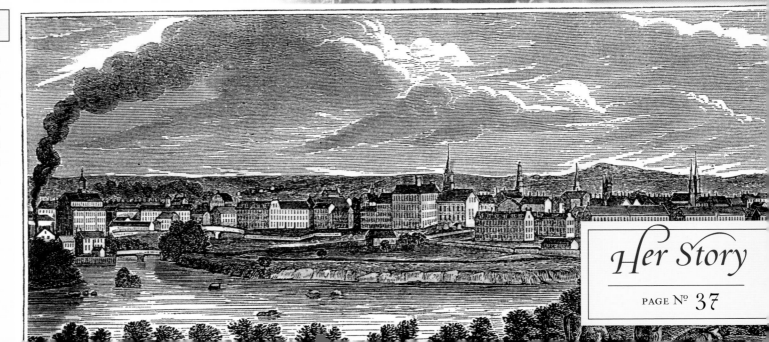

Hymn for the Working Children*

[Tune, "Autumn," or Austrian National Hymn.]

There's a voice that now is calling,
Loudly calling, day by day;
'Tis the voice of right and justice,
And its tones we must obey.
We must hasten to the rescue
Of the children young and frail,
Who are weary of their burdens,
And too soon their strength will fail.

In our stores and shops we find them,
'Mid the bloom of early spring;
But the Lord is watching o'er them,
And their calls to Him we bring.
Though their parents bid them labor
And deny their needed rest,
Yet our faith believes the promise,
That their wrongs will be redressed.

Men of rank and high position,
Men who guard our native land,
In the name of our Redeemer,
Come and lend a helping hand.
Come at once; the plea is urgent,
And the hours are waning still;
Make these children glad and happy,
And the law of love fulfil.

FANNY J. CROSBY.

1844

Sarah Bagley, a worker in a cotton mill in Lowell, Massachusetts, founds the Lowell Female Labor Reform Association to protest working conditions at the mills and to advocate the need for a shorter work day.

Her Story

PAGE № 37

1848

1848

Astronomer *Maria Mitchell* is the first woman elected to the American Association for the Advancement of Science; later she becomes the first female professor at Vassar College. Mitchell said: "In my younger days when I was pained by half educated, loose and inaccurate ways which we all had, I used to say 'How much women need exact science.' But since I have known some workers in science who were not always true to the teaching of nature, who have loved self more than science, I have said 'How much science needs women.'"

1848

1848

Elizabeth Cady Stanton helps organize a women's rights convention in Seneca Falls, New York. The convention issues the Declaration of Sentiments, containing eighteen legal grievances. This begins a seventy-two-year fight for women's right to vote. She said, "Because man and woman are the complement of one another, we need woman's thought in national affairs to make a safe and stable government."

Elizabeth Cady Stanton and her daughter, Harriot. from a daguerreotype 1856.

Frances (Fanny) Bond Palmer is the first woman to gain recognition as a lithographer.

1849

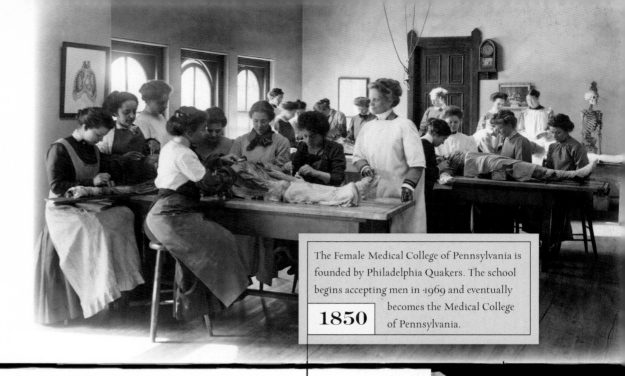

The Female Medical College of Pennsylvania is founded by Philadelphia Quakers. The school begins accepting men in 1969 and eventually becomes the Medical College of Pennsylvania.

1850

1849

Women make up 13 percent of paid workers.

1850

1850

1849

Physician *Elizabeth Blackwell* is the first U.S. woman to receive an M.D. degree when she graduates from Geneva College in New York (after being rejected because of her gender by every school in Philadelphia and New York as well as Harvard, Yale, and Bowdoin).

1850

Lucy Stone is a suffragist who organizes the first national women's rights convention; later she is among the most prominent women to keep her maiden name. She said, "And do not tell us before we are born even, that our province is to cook dinners, darn stockings and sew buttons. We want rights."

LUCY STONE

Her Story

PAGE № 39

Harriet Tubman (1823–1913)
nurse, spy and scout

Abolitionist and educator *Myrtilla Miner* overcomes serious opposition to establish and maintain the Miner School for Free Colored Girls. Miner, though white, spends all of her professional career helping educate African American women. Contributions from Quakers and from Harriet Beecher Stowe, who gives a portion of the royalties from *Uncle Tom's Cabin*, help fund Miner's work. In 1879, the Miner Normal School, a teacher training institution, becomes part of the District of Columbia's public school system. In 1929, it becomes Miner Teachers College.

1851

1851

1851

Suffragist *Amelia Bloomer* promotes dress reform for women. She advocates for a shorter, less restrictive skirt and pantaloons instead of heavy hoops and stays.

1850

1850

Harriet Tubman leads slaves to freedom via the Underground Railroad, a system of people willing to aid escapees. She said, "I had reasoned this out in my mind, there were two things I had a right to: liberty and death. If I should not have one, I would have the other, for no man should take me alive."

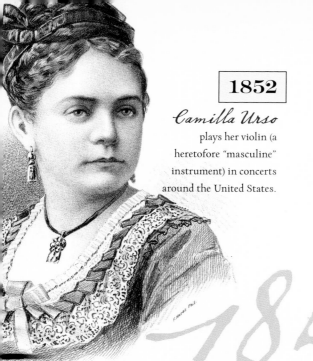

Camilla Urso
plays her violin (a heretofore "masculine" instrument) in concerts around the United States.

Abolitionist and author *Harriet Beecher Stowe* is galvanized into action by the enactment of the Fugitive Slave Act. Her best-selling book *Uncle Tom's Cabin*, which sells three hundred thousand copies in the first year, shocks the country with its frank exposé of the harsh conditions of slavery. Upon meeting her after the onset of the Civil War, President Abraham Lincoln is rumored to have said, "So, you are the little lady that started this great war."

Catharine Beecher founds the American Woman's Educational Association to recruit and train teachers to staff frontier schools. She inspires the founding of several women's colleges in the Midwest, and her writings do much to introduce domestic science into the American school curriculum. She is a strong advocate for physical education for women, incorporating calisthenics into physical education courses in the first school she establishes, Hartford Female Seminary in Hartford, Connecticut.

Emily Dickinson, who becomes a major literary figure, publishes her first poem in the local newspaper. Some critics argue that she is the greatest woman poet in the English language. Her work is studied today both in high school and on the university level. During her lifetime, she is known as a private, reclusive person who lives quietly in Amherst, Massachusetts. At the time of her death, only seven of her thousands of poems had been published. A stanza of one of her most famous poems reads:

> *Because I could not stop for Death–*
> *He kindly stopped for me–*
> *The Carriage held but just Ourselves–*
> *And Immortality.*

Suffragist
Matilda Joslyn Gage
enters the women's rights movement as a speaker. She is highly respected for her thorough organizing and her writing. She joins Elizabeth Cady Stanton and Susan B. Anthony to produce the first three volumes of the monumental *History of Women's Suffrage*. Her work for the cause is tireless; at one point she corresponds with people in forty-seven counties and circulates 2,500 tracts. Carved on her tombstone is her lifelong motto: "There is a word sweeter than Mother, Home, or Heaven; that word is Liberty."

1853

Antoinette Brown Blackwell
becomes the first formally appointed woman minister
in the United States; she is selected as the pastor of a
Congregational church.

Lawyer, teacher, and abolitionist
Mary Ann Shadd Cary issues
the first edition of a newspaper edited by a black woman
when she helps to found the *Provincial Freeman*, a
nonsectarian, nonpartisan paper in Chatham, Ontario,
Canada, where many black refugees flee.

1853

Elizabeth Keckley
buys herself freedom from
slavery by becoming an expert
dressmaker. She later becomes
the dressmaker to First Lady
Mary Todd Lincoln.

1855

1853   1854   1855   1856

1853

Harriet Goodhue Hosmer
is the first successful woman sculptor in the
United States.

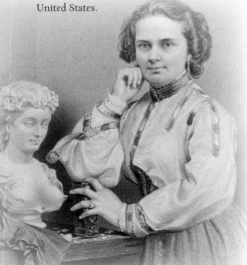

1855

Physician *Emeline Horton Cleveland*
practices gynecology and breaks down barriers against
women physicians. One of her many accomplishments is
performing operations to remove ovarian tumors; these
are judged to be the first
recorded instances of major
surgery performed by a
woman. She also establishes
training programs for nurse's
aides. Because leading male
physicians of the day consult
with her, she helps to break
down prejudices against
women doctors.

1855

Editor and journalist *Anna Elizabeth
McDowell* publishes a Philadelphia
newspaper, the *Woman's Advocate*, wholly operated
by women, where women do every job from
typesetting to printing and are paid the same as men.

During her long life, white educator, abolitionist, and suffragist *Emily Howland* supports education for blacks. She takes an interest in women's education and, eventually, makes regular contributions to over thirty schools, the majority engaged in industrial training for blacks in the South. At age ninety-nine, Howland receives an honorary doctorate from the State University of New York for her services to education.

1857

Frances E. W. Harper, a prominent black author and poet who speaks and writes about the cruelty of slavery and the lack of education and equal rights for former slaves, writes a letter to condemned prisoner and abolitionist John Brown that is published and is widely read by Northern sympathizers. Her letter says, in part: "In the name of the young girl sold from the warm clasp of a mother's arms to the clutches of a libertine or profligate,–in the name of the slave mother, her heart rocked to and fro by the agony of her mournful separations,–I thank you, that you have been brave enough to reach out your hands to the crushed and blighted of my race." At a time when most blacks cannot read and little money is available for such things as books, one of Harper's slim volumes of poetry sells more than twelve thousand copies.

1860

−1857   −1858   −1859   −1860

1859

1860

Harriet Wilson publishes a powerful early novel about the slavery experience: *Our Nig, or, Sketches from the Life of a Free Black.* This fictional autobiography reflects the conventions of both contemporary slavery narratives and the sentimentalism found in the works of many women writers of the time. A passage from *Our Nig* reads: "Her mistress entered one day, and finding her seated, commanded her to go to work. 'I am sick,' replied Frado, rising and walking slowly to her unfinished task, 'and cannot stand long, I feel so bad.' Angry that she should venture a reply to her command, she suddenly inflicted a blow which lay the tottering girl prostrate on the floor. Excited by so much indulgence of a dangerous passion, she seemed left to unrestrained malice; and snatching a towel, stuffed the mouth of the sufferer, and beat her cruelly. Frado hoped she would end her misery by whipping her to death. She bore it with the hope of a martyr, that her misery would soon close."

Businesswoman *Ellen Curtis Demorest* becomes the first person to create and distribute accurate patterns for home dressmaking.

Her Story

Physician *Mary Edwards Walker* finds women's clothes too confining. She wears men's attire and is unfazed by belittlement. She later becomes the first woman to receive the Congressional Medal of Honor for her medical services in the Civil War.

1860

1860

Educator *Elizabeth Peabody* opens the nation's first kindergarten in Boston.

In San Francisco, 85 percent of the city's population of Chinese women are enslaved as prostitutes and displayed for sale in closely guarded rooms. The practice will continue until after the turn of the century; at its peak, more than two thousand Chinese women will be sex slaves.

1860

A Chinese Slave Girl. Chinatown, San Francisco.

1860

The population in the southern states includes a quarter of a million free blacks and four million slaves.

1860

1861 The Civil War begins.

1860

Ellen White and her husband co-found the Seventh Day Adventist Church. The scriptural interpretations she receives as part of a series of more than two thousand visions are promptly accepted into church doctrine and ultimately are over nine volumes long. Her views on health, especially her opposition to coffee, tea, meat, and drugs, are incorporated into Seventh Day Adventist practices.

Mary Bickerdyke ministers to sick and wounded Union soldiers during the Civil War.

1862

For her extensive work in helping wounded soldiers, *Sally Tompkins* becomes the first woman to be a commissioned officer in the Confederate Army.

1861

1862

Physician *Marie Elizabeth Zakrzewska* founds the New England Hospital for Women and Children in Massachusetts. The hospital becomes a primary training hospital for several generations of women physicians, and nurses as well. It continues in operation today as the Dimock Community Health Center.

1861

1862

Educator *Mary Jane Patterson* becomes the first black woman to receive a bachelor's degree from an established American college when she graduates from Oberlin College.

1862

1861

In what would become West Virginia, seventeen-year-old *Belle Boyd,* passionate for the Southern cause, raises money to arm the Confederate soldiers. She wants to be more actively involved, and when her town is occupied by the Union, she beguiles military information from admiring Union officers. She is an outstanding horsewoman and uses her knowledge of the Shenandoah Valley to ferry information and act as a courier. It is believed she also smuggled needed supplies, such as quinine. She is discovered as a Confederate spy and is captured on several occasions and punished. After the war, she tells of her exploits.

1862

During the Civil War, black educator *Charlotte Forten Grimké* teaches illiterate southern slaves to read.

1862

During the Civil War, *Julia Ward Howe*, a writer of poetry, drama, and occasional travel articles for the *Atlantic Monthly* and who is active in the suffrage movement and a founder of the New England Women's Club, publishes her poem "The Battle Hymn of the Republic." Later it is set to music and becomes very popular. President Lincoln is said to have wept upon hearing it sung.

1862

Jennie Douglas becomes the first female employee at the U.S. Treasury Department. She is hired to cut and trim paper by hand, a job previously done by men. She is paid less than men for this job.

NEW-ENGLAND

FEMALE MEDICAL COLLEGE.

COLLEGE BUILDING, SPRINGFIELD STREET.

Rebecca Lee Crumpler becomes the first black woman doctor to earn an M.D. when she graduates from the New England Female Medical College.

1864

Eliza Wood Burhans Farnham publishes her most significant work, *Woman and Her Era*, which espouses the superiority of the female sex.

1864

1862

1863

1864

1864

Mary Rice Livermore organizes nurses, food, and supplies during the Civil War. It is the first time that women nurses staff army hospitals.

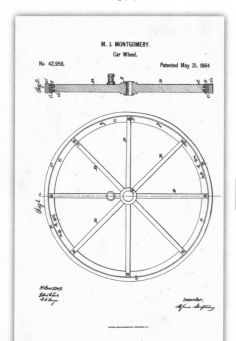

M. J. MONTGOMERY.
Car Wheel.

No. 42,958.   Patented May 31, 1864.

1864

Inveterate inventor *Mary Jane Montgomery* receives a patent for improved locomotive wheels. She is later termed the only professional woman inventor in the country by *Scientific American*.

Pauline Cushman gains the rank of major in the Union Army during the Civil War for her courage in fighting behind enemy lines.

1865

Mary Mapes Dodge publishes the popular novel *Hans Brinker, or, The Silver Skates;* more than one hundred editions of the work appear during her lifetime. The legend of the boy who puts his finger in the dike to prevent flooding is contained in the novel.

1865

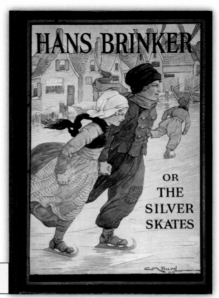

Lucy Taylor becomes the first American woman to earn a dental degree when she graduates from the Ohio College of Dentistry.

1866

1865

1866

1867

1866

Vinnie Ream Hoxie becomes the first woman to receive a federal commission for sculpture. One of her most famous works, displayed in the U.S. Capitol rotunda, is a life-size statue of Abraham Lincoln. Vinita, Oklahoma, is named in her honor.

1867

Edmonia Lewis, of African American and Native American heritage, attracts much notice with her sculpture "Forever Free" (now at Howard University). It depicts two figures, male and female, rejoicing after hearing of the end of slavery.

Her Story

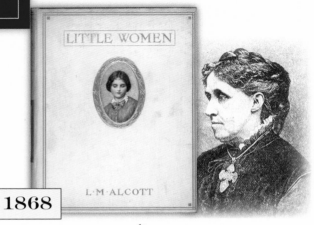

1868

Louisa May Alcott becomes widely known for her best-selling novel *Little Women*.

1868

Myra Bradwell publishes the *Chicago Legal News* despite her rejection from the bar because she is a woman. The U. S. Supreme Court said: "The natural and proper timidity and delicacy which belongs to the female sex evidently unfits it for many occupations of civil life. . . . The paramount destiny and mission of women are to fulfill the noble and benign office of wife and mother. This is the law of the Creator."

1868

1868

Mary Cassatt is the only American painter to win acceptance into the elite circle of French Impressionist painters.

1869

Suffragist and tireless speaker and writer *Susan B. Anthony* co-founds the National Women's Suffrage Association. In 1878, she persuades Senator Aaron Sargent to present an amendment giving women the right to vote; this amendment is reintroduced every year until it becomes the Nineteenth Amendment. Her strongly felt convictions include "Men, their rights and nothing more; women their rights and nothing less." Her last public words reflect her dedication to women's suffrage: "Failure is impossible."

1869

An outstanding pioneer in teacher education, *Fanny Jackson Coppin* is the first black woman to head an institution of higher learning in the United States (now Cheyney University). Part of her legacy is Coppin University.

1869

Educator and lawyer *Arabella Mansfield* is the first woman admitted to the bar in the United States. She does not practice but teaches at the university level.

Her Story

The entire United States is connected by rail when the transcontinental railroad is completed. This event significantly affects the social, political, and economic lives of women and men across the nation.

1869

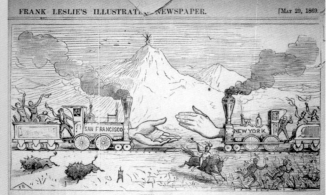

FRANK LESLIE'S ILLUSTRATED NEWSPAPER.   [MAY 29, 1869.

SAN FRANCISCO   NEW YORK

"DOES NOT SUCH A MEETING MAKE AMENDS?"

1869

1870

1870   Middle- and upper-class women often take more than two hours and the help of an assistant to dress and put up their hair. Much of this time is devoted to arranging seven to ten pounds of underwear, including corsets. Women's waists are nipped in at least four inches from their natural size by tightening metal and whalebone stays and laces and exerting anywhere from twenty-five to eighty pounds of pressure per square inch on the body. Corsets permanently alter the location of women's organs, such as stomachs and livers, and compress their ribs. Little girls' bodies are not allowed to develop normally, as they begin to wear corsets at age five.

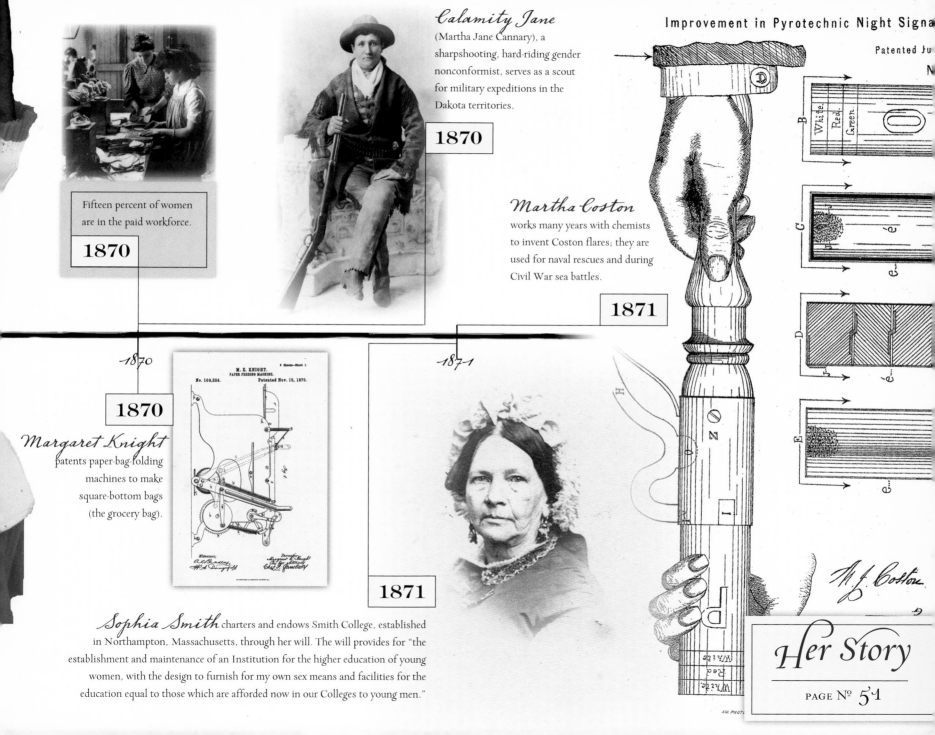

Calamity Jane
(Martha Jane Cannary), a sharpshooting, hard-riding gender nonconformist, serves as a scout for military expeditions in the Dakota territories.

1870

Fifteen percent of women are in the paid workforce.

1870

Martha Coston
works many years with chemists to invent Coston flares; they are used for naval rescues and during Civil War sea battles.

1871

1870

M. E. KNIGHT.
PAPER FEEDING MACHINE.
No. 109,224.   Patented Nov. 15, 1870.

1871

1870

Margaret Knight
patents paper-bag-folding machines to make square-bottom bags (the grocery bag).

1871

Sophia Smith charters and endows Smith College, established in Northampton, Massachusetts, through her will. The will provides for "the establishment and maintenance of an Institution for the higher education of young women, with the design to furnish for my own sex means and facilities for the education equal to those which are afforded now in our Colleges to young men."

Her Story

PAGE Nº 54

Chinese immigrant *Polly Bemis* survives being sold into slavery to become the owner of a boardinghouse and tavern; she is brought to Idaho by her owner.

1872

Charlotte Ray is the first black woman to graduate from an American law school when she receives her degree, Phi Beta Kappa, from Howard University's school of law. She becomes the first black woman lawyer to practice in the United States when she opens an office in Washington, D.C. She is also the first woman admitted to the bar in the District of Columbia.

1872

The Comstock Act is passed, banning the sale or mailing of information declared obscene, including information on birth control.

1873

1872

1873

1872

Enterprising freethinker, spiritualist, stockbroker, and later publisher *Victoria Woodhull* sets up an equal rights party and, as its nominee, runs for president of the United States.

1872

Physician *Mary Putnam Jacobi* founds the Association for Advancement of the Medical Education of Women.

1874

Mary Ewing Outerbridge brings the British game of tennis to the United States.

1873

Inventor *Helen Augusta Blanchard* invents a surgical needle, a corset-cord fastener, and a pencil sharpener, among other items, during her long career. She is best known for her sewing machine inventions, including the zigzagging sewing machine, one of her twenty-eight patents.

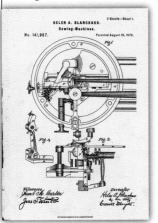

1874

1874 The Women's Christian Temperance Union (WCTU) becomes the largest American women's organization of the nineteenth century. While the early focus is on the dangers of alcohol, the issues are linked to suffrage because men are holding key political meetings in bars and saloons, to which women are denied entry. Later, the WCTU begins lobbying as a means of solving other societal issues. The WCTU becomes the oldest nonsectarian women's organization in continuous existence. Through this organization, many women learn the power of working together and speaking in public.

1874

Her Story

1874

Frances Willard is a pivotal figure in the Women's Christian Temperance Union (WCTU) as well as an activist for women's rights. She says: "I would not waste my life in friction when it could be turned into momentum."

Lydia Estes Pinkham makes "Pink Pills" for women to help with their well-being at stressful times. The pills, a patent medicine vegetable compound, are almost 20 percent alcohol, which she says acts as a preservative. Pinkham is an outstanding marketer, and her picture is on the label of all of her products, leading some to remark that hers may be the best-known American female face of the nineteenth century.

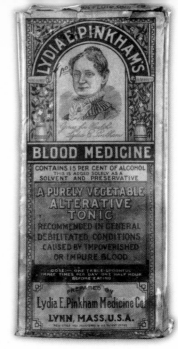

1875

-1874-   -1875-   -1876-

1875

The U.S. Supreme Court rules in *Minor v. Happersett* that *Virginia Minor*, or any woman, does not have the right to vote in Missouri. The Court rules: "Being unanimously of the opinion that the Constitution of the United States does not confer the right of suffrage upon any one, and that the constitutions and laws of the several States which commit that important trust to men alone are not necessarily void . . ."

1876

The only woman naturalist of her time and the first taxidermist to place specimens in realistic poses and settings, *Martha Maxwell* exhibits stuffed birds at the Centennial Exhibition in Philadelphia.

1876

Colorado grants women partial suffrage in local school elections.

1877

1878

1878

Clara Shortridge Foltz is admitted to the bar in California and becomes that state's first woman lawyer. She works to pass legislation allowing women to practice law in the state and is the first admitted under its provisions. Although denied admission to the Hastings College of Law, she sues, argues her own case, and wins. Hastings' appeal to the California Supreme Court is denied (*Clara Foltz v. J. P. Hoge et al.*). Women are allowed to enter and graduate from Hastings as a result.

1879

1879

Mary Baker Eddy establishes the Christian Science Church.

Her Story

PAGE Nº 55

1879

Belva Lockwood lobbies Congress to pass a bill admitting female attorneys to practice before the U.S. Supreme Court. In 1884, the National Equal Rights Party nominates her for U.S. president.

1881

Clara Barton, with the help of a handful of friends, establishes the American branch of the Red Cross and becomes its first president, serving in this position for twenty-three years. During that time, the organization becomes widely recognized for providing disaster relief, both foreign and domestic, in more than twenty-one instances.

1880

Susette La Flesche Tibbles, a member of the Omaha tribe, serves as an interpreter, lecturer on Indian affairs, and spokesperson. She contributes to the passage of the Dawes Act, which authorizes the allotment of reservation land to individual Native Americans.

1879   *1880*   *1881*

1879

Ida Lewis is the first woman to become an official U.S. lighthouse keeper.

1880

Minister, lecturer, and suffragette *Anna Howard Shaw* is ordained as a minister in the Methodist Protestant Church. Later, while in medical school, she becomes active in the suffrage movement. Shaw works for many years in the leadership of the movement, living long enough to know that the Nineteenth Amendment has passed both houses of Congress and is on its way to ratification.

1880

Women join the Knights of Labor as valued members; when the organization declines in the 1890s, there is no national labor association that accepts women.

1882

Elizabeth Agassiz founds the Society for the Collegiate Instruction of Women, which in 1894 will become Radcliffe College. She also serves as the school's first president.

1882

1882

Actress *Minnie Maddern Fiske*, who began her career at age three with occasional roles in a troupe managed by her father, makes her stage debut in New York at age sixteen. During the next two decades, her career continues to flourish and she writes plays as well as acts. At the height of her career, she is hailed as the chief American representative of naturalism in the theater.

1882

Chemist *Ellen Swallow Richards*, historian and educator Alice Palmer, and others found the Association of Collegiate Alumnae, which would later be called the American Association of University Women. The AAUW is still active today.

1882

Mathematician *Christine Ladd-Franklin* finishes her Ph.D. degree at Johns Hopkins, but it is not awarded because she is a woman. She finally receives it in 1926.

1883

Sarah Winnemucca is a Paiute Indian who gathers signatures demanding the land grants promised to her people.

1883 The first "Harvey Girls" go west to work as waitresses in Fred Harvey's restaurants along the Santa Fe Railroad lines. They are required to commit to a minimum six-month contract of work and travel (during which time they may not marry).

1883

1883

Emma Lazarus writes a sonnet, entitled "The New Colossus," that is chosen to be inscribed on the base of the Statue of Liberty.

The Civil Service Act is passed, enabling women to compete with men for government jobs. **1883**

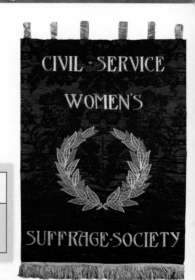

CIVIL · SERVICE

WOMEN'S

SUFFRAGE · SOCIETY

1884

Painter *Cecilia Beaux,* considered to be the finest woman painter active in America at the turn of the century, completes her first important artistic work. One of the top portrait painters in the United States, Beaux wins prizes in museum exhibitions from Philadelphia to New York to Paris. In 1895, she becomes the first full-time woman faculty member at the Pennsylvania Academy of the Fine Arts, where she teaches for twenty years.

1885

Destined to become America's great master of the violin, *Maud Powell* makes her American debut with the New York Philharmonic Orchestra. She revolutionizes the art of violin playing with her first records in 1904, when she is selected by the Victor Company as its first solo instrumentalist for its new artist series. Among the supreme violinists of the time, Powell is also a popular favorite who makes it her mission to bring the best in classical music to all Americans, both in remote areas and in large cities.

1884

1884

In the dress of their times, women carry out a myriad of duties, including farm chores and branding cattle.

SCENE AT A SAN LOUIS VALLEY CATTLE RANCH

1885

1885

Sharpshooter *Annie Oakley* (Phoebe Ann Mosey) joins Buffalo Bill's Wild West Show.

Her Story

PAGE № 59

1886

Philanthropist *Grace Hoadley Dodge* contributes time, energy and money to many worthwhile social causes. Her efforts lead to the foundation of a teachers college in New York City, which in 1900 becomes part of Columbia University.

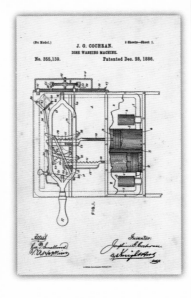

1886

Josephine Cochran invents the first practical dishwasher.

1886

Chicago labor activist and organizer *Lucy Parsons* (born Lucia González), with her husband, leads eighty thousand striking workers up Michigan Avenue as part of a protest to demand an eight-hour workday, in what becomes known as the Haymarket Affair. Parsons, who probably was born a slave and is of African American, Native American, and Mexican American heritage, fights against poverty and social injustice all of her life. She is a prominent lecturer, and as a woman of color who stands against racism and sexism, she is often arrested before she even speaks.

—1885

1885

Sarah Goode becomes the first black woman to receive a U.S. patent for her folding cabinet bed, what we today call a "hide-away." When it is not used as a bed, it serves as a desk.

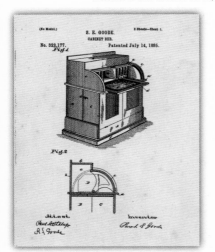

—1886

1886

Sophia Hayden Bennett becomes the first woman admitted to the Massachusetts Institute of Technology to study architecture.

—1887

Anne Sullivan Macy (at right) begins her career as the governess to a deaf and blind child, Helen Keller.

1887

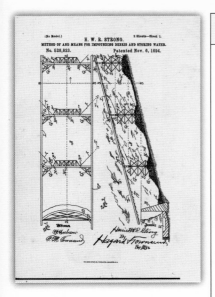

Businesswoman *Harriet Strong* invents water conservation and irrigation techniques that promote the development of southern California as a major agricultural region.

1888

Franciscan nun *Mother Marianne Cope* moves to Kalaupapa, Hawaii to help at an outcast leper colony. When famed priest Father Damien de Veuster dies of leprosy, she takes over the care of the patients. Amazingly, she is not infected by this highly contagious disease during her many years of close contact. She is beatified in 2005, one of the steps toward sainthood in the Roman Catholic Church.

1889

Anna Bissell introduces progressive labor policies at the Bissell carpet sweeper company, as president and chair of the largest organization of its kind in the world.

-1888

1887

Frances Wisebart Jacobs organizes the Charity Organization Society, which is the predecessor of the United Way.

1889

-1889

Painter and poet *Lilla Cabot Perry* encourages intellectual discussion of the ideas of the day when she holds informal salons in her home. She is among the first American artists of the late nineteenth century to paint in the Impressionist style. Her artistic work is influenced by her friend and mentor Claude Monet.

1888

Sissieretta Jones makes her singing debut; her troupe, the Black Patti Troubadours, tours from 1896 to 1916.

1889

Nellie Bly
(Elizabeth Cochrane Seaman) is an investigative reporter who goes around the world in seventy-two days.

1889

Jane Addams begins Hull House, a neighborhood community center offering a full range of health and social services. For this achievement, she receives the Nobel Peace Prize in 1931. She will second Theodore Roosevelt's nomination for president in 1912. Newspapers call her "one of the ten greatest citizens of the Republic." Hull House becomes a model of neighborhood social support, which is replicated in other cities, and still operates to this day.

1889

1889

Writer *Kate Chopin* publishes her first two stories. She goes on to publish more than one hundred well-polished short stories. Chopin becomes known for writing about the oppression of married women, and her novel *The Awakening* centers on a character who rejects the traditional roles of married women of the day.

1889 Photographer and photojournalist *Frances Benjamin Johnston* is the official White House photographer for five presidential administrations: those of Benjamin Harrison, Grover Cleveland, William McKinley, Theodore Roosevelt, and William Howard Taft. She gains further renown when she is commissioned to document the success of the first higher educational institution to admit both African Americans and Native Americans, what would later become Hampton University in Hampton, Virginia. Her first camera is a gift from George Eastman, of Eastman Kodak fame. More than twenty thousand of her photographs are housed in the Frances Benjamin Johnston collection at the Library of Congress.

1889 *Susan La Flesche Picotte* graduates from medical school at the Women's Medical College of Pennsylvania at the top of her class. She becomes the first Native American woman physician in the United States to earn a medical degree. She returns to the Omaha reservation in Nebraska, where she serves as a physician and works to improve health conditions of the tribe.

Jane Cunningham Croly **1889** founds and becomes the first president of the Women's Press Club of New York.

1889 Roman Catholic nun *Mother Frances Xavier Cabrini,* who from childhood wanted to make religious life her work, comes to the United States to work among poor Italian immigrants. In 1946, she becomes the first American citizen to be made a Roman Catholic saint.

1889 Lifelong radical and activist *Emma Goldman* immigrates to New York City; later she says that her life began on this date. She agitates on behalf of unrestricted liberty; eventually she co-founds the radical journal *Mother Earth*. Goldman is also very pivotal in the movement supporting birth control and speaks in public forums to attract support for the cause.

Her Story

PAGE № 63

Prominent feminist and reformer *Alice Stone Blackwell* (daughter of Lucy Stone) leads the movement to reconcile the two competing factions of the woman's suffrage movement. The differences between the groups are based on differences in strategy: one favors state-by-state campaigns, the other a federal constitutional amendment. Stone Blackwell helps the National American Woman Suffrage Association (NAWSA) combine both of these techniques in order to secure the passage of the Nineteenth Amendment. When suffrage becomes a reality, NAWSA reorganizes as the League of Women Voters.

1890

1890

Black educator and former slave *Lucy Craft Laney* founds a teachers college called the Haines Normal Institute in Atlanta, Georgia.

1890

Kate Gleason, who served as bookkeeper for the family business from the age of fourteen and became the first woman to enter Cornell University's engineering program, is called back home by her father to work for the family company. Becoming secretary-treasurer, she markets gear-cutting machinery internationally and helps the firm grow into a nationally prominent producer. After leaving the company, she is appointed receiver of a bankrupt company in 1914, probably the first woman in the United States to be so named, and in 1918 becomes a bank president. In 1998, the Rochester Institute of Technology names the Kate Gleason College of Engineering in her honor. The Gleason Corporation is still in existence today in Rochester, New York.

1890

Ida Gray Nelson becomes the first black woman to earn a dental degree.

1890

Daughters of the American Revolution (DAR) is formed; it is a group that honors the heritage of Americans who can trace their lineage in this country back to the late 1700s.

1890

Rose Markward Knox takes over a gelatine business upon the death of her husband and runs it for more than forty years, during which time it becomes the largest such company in the United States. She is still chair of the board of directors at the time of her death. Knox is the first woman to be elected director of the American Grocery Manufacturers Association.

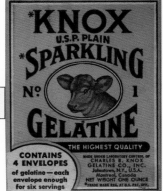

*KNOX
U.S.P. PLAIN
*SPARKLING
NO 1
GELATINE
THE HIGHEST QUALITY
CONTAINS 4 ENVELOPES
of gelatine—each envelope enough for six servings
MADE UNDER LABORATORY CONTROL OF
CHARLES B. KNOX
GELATINE CO., INC.
Johnstown, N.Y., U.S.A.
Montreal, Canada
NET WEIGHT ONE OUNCE
*TRADE MARK REG. AT U.S. PAT. OFF.

1891

Martha Matilda Harper develops the modern franchising system in her chain of skin and hair care salons. She uses a picture of herself with her long hair as part of her advertising.

1891

Mary Emma Woolley becomes the first woman admitted to Brown University. She will also become the first woman Phi Beta Kappa and the first woman to represent the United States at a major diplomatic conference. She later becomes president of Mount Holyoke College.

1892

Author, educator, and scholar **Anna Julia Cooper,** born a slave but the fourth African American woman to earn a Ph.D. degree, publishes *A Voice from the South: By a Woman from the South*, an early example of black feminism. Her life is dedicated to educating fellow blacks. A traffic circle in Washington, D.C., is named in her memory.

1891

1892

1891

Sister Mary Katharine Drexel founds the religious order Sisters of the Blessed Sacrament for Indians and Colored People in the Philadelphia, Pennsylvania, area. She is canonized as a Roman Catholic saint in 2000.

1891

Author, folklorist, and defender of Native American rights **Harriet Maxwell Converse** is the first Anglo woman to become an honorary chief of a Native American tribe, the Seneca Nation.

1892

Feminist and writer **Charlotte Perkins Gilman** publishes *The Yellow Wallpaper*, about a woman driven mad in her repressive marriage. Gilman becomes the leading feminist theoretician of her generation.

Her Story

1892

Bertha Palmer is selected as president of the Board of Lady Managers for the World's Columbian Exposition in Chicago. Palmer and the board organize the Women's Building.

1893

Florence Bascom, pioneering geologist, becomes the first woman to receive a Ph.D. from Johns Hopkins University. She is the first woman hired by the U.S. Geological Survey.

1892

1892

Mary Kenney O'Sullivan is appointed by Samuel Gompers as the first woman general union organizer.

1893

1892

Ellis Island opens for immigration screening. Immigrants undergo inspections to determine if they have medical conditions or legal problems that would preclude their admission to the country.

1893

Hannah G. Solomon founds the National Council of Jewish Women, an important charitable and philanthropic organization.

1894

Josephine St. Pierre Ruffin organizes the Woman's Era Club, among the first of the Negro women's civic associations. Ruffin says: "It is the women of America–black and white–who are to solve this race problem, and we do not ignore the duty of black women in the matter. They must arouse, educate, and advance themselves. The white woman has a duty in the matter also. She must no longer consent to be passive. We call upon her to take her stand."

-1894

1894

Born into the American theater's "royal" family, actress *Ethel Barrymore* makes her stage debut at age fourteen in *The Rivals*. She earns her position of first lady of the American stage over the course of a career that includes appearances on the stage and in film. She wins an Academy Award and establishes the Ethel Barrymore Theatre in New York City, still in operation today.

1893

Annie Laurie (Winifred Black) is a fearless reporter who goes to any lengths to get her story. She later disguises herself as a boy to report on the Galveston, Texas, flood of 1900.

1895

Katherine Lee Bates

publishes her poem "America the Beautiful," which becomes the basis for a widely beloved patriotic song.

Activist, suffragist, and integrationist *Ida B. Wells-Barnett* publishes *A Red Record*–a detailed look at lynching. She speaks and writes about these crimes throughout her long life. In *Crusade for Justice*, she says, "One had better die fighting against injustice than die like a dog or a rat in a trap."

1895

Botanist *Catherine Furbish,* who was studying local plants by the time she was twelve and who spent the next thirty-five years collecting, classifying, and recording the flora of Maine, founds the Josselyn Botanical Society in that state. Although she is an amateur, contemporary professional botanists recognize the quality of her work. She donates her large folio volumes to Bowdoin College and her dried plant collection to the New England Botanical Club.

1895

As a teenager, *Catherine Evans Whitener* masters the candlewicking technique that she saw in a bedspread that was a family heirloom. The tufted quilts and bedspreads she creates, often referred to as chenille, as well as mats and bathrobes form the basis of the Evans Manufacturing Company, which she founds in 1917 with her brother. By 1941, the bedspread industry, centered in Dalton, Georgia, employs ten thousand people and has sales of over $25 million. In the twenty-first century, more than 90 percent of machined carpet production is tufted, based on Whitener's work. She remembers: "When I was a girl I wished that I had been a boy. Because a boy could find work to make money, and there was nothing a girl could do to earn money. I feel now that God knew best, and I am glad that I was a girl."

1895

Mary Engle Pennington is denied a B.S. degree at the University of Pennsylvania because of her gender. She is not deterred and continues her academic work, later earning a Ph.D (conferred under an old statute that makes exceptions for female students in extraordinary cases) and becoming a bacteriologist and chemist. Pennington's research on bacteria and refrigeration leads to safer eggs, poultry, and fish. She is the first female member of the American Society of Refrigeration Engineers.

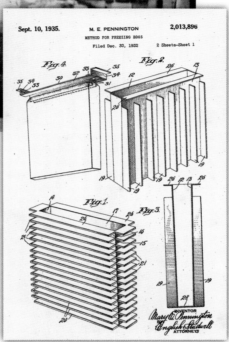

Sept. 10, 1935.   M. E. PENNINGTON   2,013,896
METHOD FOR FREEZING EGGS
Filed Dec. 30, 1932   2 Sheets-Sheet 1

1895

Social reformer *Lillian D. Wald,* who comes from an affluent family and is deeply troubled by the inadequacy of the health facilities available for new immigrants, establishes the Nurses Settlement House at 265 Henry Street, on New York's Lower East Side, and coordinates nursing classes. Later called the Henry Street Settlement House, the organization helps to improve life in immigrant ghettos.

1895

Hopi potter *Nampeyo* designs and creates vases and vessels that are inspired by earlier Native American art and her own creativity.

Amy Marcy Cheney Beach, who was composing piano tunes by age four and made her concert debut as a pianist at age sixteen, becomes the first woman composer to have one of her symphonic musical works performed when the Boston Symphony presents her Symphony in E-minor (the "Gaelic" Symphony).

1896

Mary Parker Follett is one of the first people to apply psychological insights and social science findings to the study of industrial organizations.

1896

Fannie Farmer revolutionizes cooking with the publication of the *Boston Cooking-School Cook Book,* which contains standardized measurements in recipes for the first time.

1896

Writer *Sarah Orne Jewett,* who spends her whole life in the Maine countryside and becomes a prolific writer of short stories, sketches, and a few novels, publishes the classic *The Country of the Pointed Firs.* In 1901, Bowdoin College granted her an honorary literary doctorate; she is the first woman so honored.

1896

Astronomer *Annie Jump Cannon* is hired at the Harvard College Observatory. She later develops the scheme of stellar classification and goes on to classify over half a million stars, more than any other person had accomplished in the past.

Civil rights activist *Mary Church Terrell* forms the National Association of Colored Women. At age eighty-six, she leads the successful fight to integrate eating places in the District of Columbia.

1898

Julia Morgan, a civil engineer, becomes the first woman to study architecture at the École des Beaux-Arts in Paris (after being refused admission for two years because of her gender). Her architectural practice in San Francisco, established in 1904, is especially busy after the 1906 San Francisco earthquake. Her many projects for the YWCA include the Asilomar Conference Center in Pacific Grove, California. After completing more than 450 commissions, she is selected by William Randolph Hearst to serve as the architect for his ranch at San Simeon, California. Today, the project, on which she worked for twenty years, is called the Hearst Castle and is a California state park.

1897   *1898*

1897   *Alice McLellan Birney* (front row, third from left) organizes the National Congress of Mothers, a forerunner of the Parent Teacher Association.

Her Story

PAGE N° 74

1899

A nationally known speaker of the Women's Christian Temperance Union (WCTU), suffragist *Carry Nation* garners much publicity for her cause by carrying and using a hatchet to smash bottles in saloons, all the while singing hymns and praying. She passionately spearheads the movement with her belief that it is her divine mission to do this work. She is the topic of numerous books and articles that discuss the battles against alcohol in pre-Prohibition America.

1899

Florence Kelley (third from left) is an activist in the National Consumer's League; she is also one of the most effective social reformers at Jane Addams' Hull House.

1900

Margaret Abbott is the first American woman to win an Olympic gold medal, although she did not know the all-women nine-hole golf tournament for which she won a bowl (and in which she competed against her mother) was an Olympic event.

WINS WOMEN'S GOLF CUP.

MISS MARGARET ABBOTT LEADS PLAY NEAR PARIS.

Some Interesting Contests Decided at Compiègne Links—Gallery Interferes with Some of the Players—New Tenor Makes Successful First Appearance in Opera—Exposition Attendance Falls Off Several Thousand for the Week—Novelties at Theaters.

[Special Cable to the New York Tribune and The Chicago Tribune by C. I. Barnard.]

PARIS, Oct. 6.—There has been a falling off in the exhibition attendance, the daily average paying entrances being 240,000, as against 230,000 for last week. Numbers of well-known Americans continue to arrive, many of whom attended the international golf matches given under the auspices of the exhibition on Compiègne links. The women's championship attracted the most attention. Among those present were Mr. and Mrs. Marshall Field Jr., Mr. and Mrs. Amory Lawrence, Mr. Caleb Curtis and his two daughters, Mr. and Mrs. John E. Drexel, Mr. Drexel Paul and Miss Paul; Mrs. Whitney Warren, who is visiting her father, Mr. C. C. Tooker, Mrs. John Whitcomb Cotton, Miss Cotton, Mr. and Mrs. Edward Blair, Mrs. Levi P. Morton and daughters, Mr. and Mrs. H. O. Havemeyer, Mrs. John Sargent Cram, and a smart Parisian contingent from the Faubourg Saint Germain. There were ten entries for the women's event, among whom was Miss Margaret Abbott, who a couple of years ago belonged to the Chicago Golf club, and played with Mrs. Chatfield-Taylor and other Western players on the Wheaton links.

1899

1899

The National Association of Women Lawyers is formed.

1899

Mohawk Indian *L. Rosa Minoka-Hill,* who at five years of age went to live with a prominent Quaker physician in Philadelphia, graduates from the Woman's Medical College of Pennsylvania at age twenty-three. She provides medical care to Wisconsin's Oneida Indian tribe for more than forty years.

1900

Twenty percent of Anglo women and 40 percent of black women are in the paid workforce.

1900

Twenty-five percent of all boys from ages ten to fifteen and 10 percent of all girls between those ages work for non-farm wages.

1901

In Page County, Iowa, *Jessie Field Shambaugh* organizes before- and after-school boys' and girls' clubs addressing needs of young people in agricultural communities. These clubs evolve into the 4-H organization, symbolized by a four-leaf clover representing head, heart, hands, and health.

-1900-

1900

Carrie Chapman Catt becomes the president of the National American Woman Suffrage Association.

1900

Nannie Helen Burroughs speaks at the National Baptist Convention about the right of women to participate equally in missionary church work.

-1901-

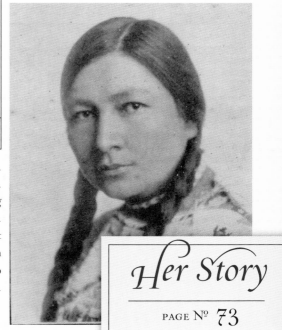

1901

Zitkala Sa (Red Bird), also known as Gertrude Bonnin, writes *Old Indian Legends* drawing on her Yankton Sioux heritage. She overcomes significant prejudice as a Native American and as a woman to contribute to literature, music, and politics.

1901

Sophonisba Breckinridge becomes the first woman to receive a Ph.D. in political science at the University of Chicago. She becomes deeply involved in social reform and moves to Chicago's Hull House to join other women interested in these issues.

1901

Suffragist, social reformer, and historian *Maud Wood Park* discovers while in college and later at the national suffrage convention that young women do not favor suffrage and she is in a significant minority. This encourages her to travel nationwide and set up chapters of college suffrage associations in more than thirty states and also to co-found the Boston Equal Suffrage Association for Good Government in 1904. She serves as its executive secretary for twelve years. Park works to gain passage of many social reform bills and gathers materials on women's rights for the Schlesinger Library, preserving that history for all.

1902

Ethnographer and anthropologist *Alice Cunningham Fletcher* does fieldwork with the Sioux recording their tribal language and songs. She is a founding member of the American Anthropological Association.

1902

Annie Malone is going door-to-door selling her Poro products, which are hair treatments for African American women. She later establishes Poro College in St. Louis, where students learn about cosmetology.

1902

Agnes Nestor is a founder of the International Glove Workers Union. She later becomes the first woman elected president of an international labor union.

1903

At age five, *Molly Picon* wins first prize in a children's contest for musical acts. She later is one of the preeminent Yiddish-speaking actresses, singing and dancing in New York's Second Avenue Theatre for more than thirty years.

1903

Maggie Lena Walker becomes a successful banker when she founds and owns the first U.S. African American bank in Richmond, Virginia.

1902

1903

1902

Harriet Stanton Blatch (daughter of Elizabeth Cady Stanton) begins her work to revitalize the women's suffrage movement, which is faltering after fifty-four years of hard work and limited progress. She organizes women employed in factories, laundries and garment shops; she holds open-air meetings, inaugurates the first suffrage parades, sends women to testify before the New York legislature, campaigns in election districts, and stations women at the polls. She later devotes all of her efforts to helping the Allies win World War I, and spends twenty years championing the rights of workingwomen.

Mother Jones (Mary Harris), a key figure in the American labor movement since 1877, leads a caravan of child laborers from Pennsylvania to the New York home of President Theodore Roosevelt to protest the exploitation of children. This continues her longtime practice of moving from one industrial area of the country to another, organizing strikes and educating workers. In 1905, she and Lucy Parsons are among the cofounders of the Industrial Workers of the World (IWW).

1903

Her Story

1903

Helen Keller, who became blind, deaf, and mute at the age of nineteen months, publishes her autobiography, *The Story of My Life*. Through this and her other books, as well as her lectures, she serves as a courageous role model encouraging the accomplishments of others. Handicapped individuals were removed from asylums due to Keller's efforts to improve their treatment. In 1963, she is awarded the Presidential Medal of Freedom. Keller says: "When indeed shall we learn that we are all related one to the other, that we are all members of one body? Until the spirit of love for our fellowman, regardless of face, color or creed, shall fill the world, making real in our lives and our deeds the actuality of human brotherhood–until the great mass of the people shall be filled with the sense of responsibility for each other's welfare, social justice can never be attained."

1903

1903

Rose Schneiderman organizes the first women's local of the Jewish Socialist United Cloth Hat and Cap Makers Union. From 1907, she devotes much of her time to the Women's Trade Union League. A compelling speaker, she later pushes reforms such as the eight-hour workday and the minimum wage.

1904

Ida Tarbell writes about the Standard Oil Trust, exposing blackmail and price rigging.

1904

Lena Bryant, whose name is misspelled on a business account application, opens her first Lane Bryant shop in New York. Her innovative contributions to retailing include the production of the first commercial maternity dress which, for the first time, allows women to appear in public when they are pregnant. When newspapers will not accept advertising for maternity clothes, Lane Bryant opens a mail order business. She also initiates clothing lines for full-figured women. By 1950, Lane Bryant is the sixth-largest catalog retailer in the United States.

1904

Educator and political activist *Mary McLeod Bethune,* fifteenth of seventeen children born to former slaves, founds the Daytona (Florida) Educational and Industrial Training School for Negro Girls. Through several stages of evolution, it becomes a college and today is known as Bethune-Cookman College. She later forms the National Council of Negro Women. Bethune said: "Believe in yourself; learn and never stop wanting to build a better world."

—1904

—1905

1905

1905

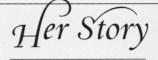

1905

Portions of the diary of *Mary Chesnut* are published posthumously. During the war, Chesnut kept a detailed diary that was an eyewitness account of the Civil War years from a Southern perspective. The account was influenced by the perspective of Chesnut's husband, Confederate General James Chesnut Jr., who was an aide to Confederate President Jefferson Davis. Chesnut's diary, when republished in 1984, wins a Pulitzer Prize the following year. The diary is described as the most important piece of literature produced by a Confederate author during that period.

The research of biologist and cytogeneticist *Nettie Stevens* contributes greatly to the understanding of chromosomes and heredity, particularly the role of the X and Y chromosomes.

Architect *Mary Colter* designs the Hopi House on the south rim of the Grand Canyon, having worked with railroad magnate Fred Harvey since 1904 to create a number of hotels and lodges to bring tourists to the southwestern United States. Her design of the Bright Angel Lodge, also on the south rim of the Grand Canyon, in 1935 leads to an architectural genre referred to as "National Park Service Rustic." This genre uses materials native to the site and large-scale design elements.

Her Story

PAGE № 77

1905

1905

The daughter of former slaves,
Madam C. J. Walker
has become the richest woman in America
through her hair and cosmetics business.
She is cited by the Guinness Book of Records
as the first female American self-made millionaire.
Walker said: "I am a woman who came from the
cotton fields of the South. From there I was promoted to the washtub. From there I was promoted
to the cook kitchen. And from there I promoted
myself into the business of manufacturing hair
goods and preparations. . . . I have built my own
factory on my own ground."

Fanny Bullock Workman is the first female mountaineer to reach an altitude of over twenty-three thousand feet (in the very long skirt of the time) when she climbs Nun Kun in the Himalayas.

1906

Williamina Stevens Fleming is the first U.S. woman elected to the Royal Astronomical Society.

1906

Singer and actress *Sophie Tucker* launches a show business career that spans sixty years.

1906

-1906

-1907

-1908

-1909

More than twenty thousand Japanese and Korean "picture brides" come to the United States through Hawaii between 1907 and 1924. They are very young and often come only with a picture of their husband-to-be. Their lives are often difficult because of very hard work, language, and culture issues.

1907

American Red Cross fund-raiser and leader in the fight against tuberculosis, *Emily Bissell* designs and prints the first Christmas seals.

1907

1908

Social reformer *Edith Abbott* begins work at Hull House; she writes more than a hundred books and articles about public welfare and social injustices.

1909

Social activist *Mary White Ovington* is one of the founders of the NAACP (National Association for the Advancement of Colored People) and during the early years one of the organization's few nonblack members.

1909

1909

1909

Gertrude Stein, a writer and woman of powerful intellect and firm opinions, publishes her first book, *Three Lives*. She is well known for her *Autobiography of Alice B. Toklas*, published in 1933, which was actually her autobiography. Alice B. Toklas was her secretary and companion from 1912. Stein's prose style is concerned with sound and word rhythms, and she likened it to abstraction or Cubism in painting. She wrote: "Rose is a rose is a rose is a rose," often quoted as "A rose is a rose is a rose."

1909

Helen Hayes makes her Broadway debut at age nine; by 1930 she is considered the first lady of American theater.

1909

Alva Erskine Smith Vanderbilt Belmont rents a floor of a New York office building as headquarters for the National American Woman Suffrage Association, for a press bureau whose activities she finances, and for the Political Equality League, of which she is founder and president. She becomes an ardent feminist and suffragist after the death of her second husband.

1910

Charlotte Vetter Gulick founds the Camp Fire Girls, an organization for young women.

Elizabeth Arden starts a chain of beauty salons selling both a line of upscale cosmetics with her brand and cosmetic services such as facials. Her Red Door salons introduce the makeover concept and are still in business today.

1910

Conquest of Colour

Harmony and variety go hand in hand if Elizabeth Ar is your guide. No danger of spoiling the perfect dr by combining it with the wrong make-up. No risk allowing some slight decorative flaw to creep into otherwise impeccable colour scheme. Miss Arden devised the right make-up for each new shade in fashic colour card for Spring—the right lipstick, rouge, eye sha eye pencil, cosmetique, nail varnish and powder for ev ensemble. Jewels, make-up, dress should form a sympho Write for Miss Arden's charts, "The Conquest of Colou

Elizabeth Arden

25 OLD BOND STREET LONDON WEST ON

PARIS: ELIZABETH ARDEN S.A.

NEW YORK: 6TH FIFTH AV

The various make-ups, excluding the powders, be had separately in a charming box at

1910

Lawyer *Crystal Eastman* drafts early worker's compensation laws. An ardent advocate of social causes, she later co-founds the American Civil Liberties Union (ACLU) and is one of four people who write the Equal Rights Amendment (ERA), introduced in 1923.

1910

Interpreter *Tye Leung Schulze* is the first Chinese American to work in the federal government.

Harriet Quimby is the first American woman and the second woman in the world to become a licensed airline pilot.

On March 25, a fire breaks out in the Triangle Shirtwaist Company factory in New York City. One hundred forty-six workers, most of them young women, are killed; after this tragic event, unions push for worker safety protections.

1911

1911

At age sixty-one, mountain climber *Annie Smith Peck* plants a "Votes for Women" pennant atop Peru's Mount Coropuna.

1911

1911

Educator *Katharine Gibbs* develops the first school for women who want to become secretaries.

1911

Actress *Mae West* makes her Broadway debut; she dominates the movie screen during the 1930s, often using the language of sexual innuendo. One of her often-quoted, well-known lines was "Why don't you come on up and see me sometime . . . when I've got nothin' on but the radio."

1911

1911

Journalists *Jovita Idar* and *Soledad Peña* organize the League of Mexican Feminists.

Elizabeth Gurley Flynn (second from left) is a labor leader who organizes the "bread and roses" strike of twenty thousand textile workers in Lawrence, Massachusetts.

1912

1912

Mary Antin publishes *The Promised Land*, her autobiography about her immigrant experience; it is used as a school civics text until 1949.

THE PROMISED LAND

BY MARY ANTIN

WITH ILLUSTRATIONS FROM PHOTOGRAPHS

Gracie Allen first appears in vaudeville; she is very quick to deliver one-liners and clever jokes. Her comedic talent is heard on radio and seen on the stage and later on television.

1912

1911

1911

Pathologist Maud Slye joins the staff of the Sprague Memorial Institute to continue her studies on the hereditary transmission of cancer.

1912

Social reformer and activist Julia Lathrop is the first woman to head a statutory federal bureau; President Taft appoints her and the Senate confirms her as the chief of the Children's Bureau.

1912

Henrietta Szold founds Hadassah, now the largest worldwide Jewish women's organization, promoting education, youth programs, and health care.

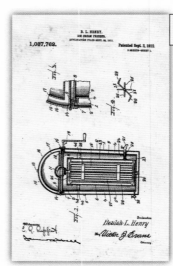

1912

Inventor *Beulah Henry,* called "Lady Edison" in the press and at the patent office, receives her first patent, for an ice cream freezer.

Juliette Gordon Low (center) organizes the first American Girl Guide troop (later called the Girl Scouts). By 1927, there will be a Girl Scout troop in every state.

19·12

1912

19-12

Anita Loos sells her first screenplay. Later her novel *Gentlemen Prefer Blondes* becomes a best seller.

Henrietta Swan Leavitt discovers the period-luminosity relation for those stars whose brightness changes (variable stars), which allows calculation of distances to far-off galaxies.

Militant activists *Lucy Burns* (above) and *Alice Paul* (left) reinvigorate the campaign for a federal suffrage amendment for women. Their public protests land them in jail on more than one occasion. A spectacular success of their collaboration is the suffrage parade in 1913, the day before Woodrow Wilson's inauguration as president.

Labor advocate *Josephine Roche* becomes Denver's first police-woman. After holding a series of federal government positions, she serves for twenty-four years as the executive director of and neutral trustee for the United Mine Workers' Welfare and Retirement Fund.

Crisscrossing the Northwest, traveling thousands of miles and delivering hundreds of lectures, *Abigail Scott Duniway* speaks on suffrage and temperance issues. At age seventy-eight, she signs Oregon's suffrage proclamation and becomes the first woman in her state to register to vote.

Reformer *Belle Moskowitz* is an impartial arbitrator for the garment workers union and for dress manufacturers.

America's first recognized interior decorator, *Elsie de Wolfe*, publishes *The House in Good Taste*, which helps shape the taste of an entire generation. While many of her clients are wealthy, her book includes numerous suggestions that are practical and inexpensive to implement.

Her Story

1914

Mary Phelps Jacob receives a patent for the "backless brassiere"; it is the first patent for an undergarment with the name *brassiere*.

19–14

1914

Helena Rubinstein opens her first beauty salon in the United States. She starts the Helena Rubinstein Foundation in 1953 to put in action her often-expressed principle: "My fortune comes from women and should benefit them and their children, to better their quality of life." The foundation provides scholarship grants to young women to encourage them to pursue their education.

RUTH SAWYER COLLECTION

1915

Ruth Sawyer publishes *The Primrose Ring*, which is made into a movie in 1917; in 1937 she receives a Newbery Medal for her novel *Roller Skates*.

Marianne Moore publishes her first poems; she has a long successful career editing *The Dial* (a famous literary magazine) and publishing poetry and literary criticism.

1915

1914 | President Wilson proclaims Mother's Day a national holiday.

—19-14—

—19-15—

Ruth St Denis is a cofounder of the first major professional dance school.

1915

1915

1915 | The Women's Peace Party is formed; by 1917 it has more than forty thousand members.

PEACE

Heralded as the "premier black pianist" for over four decades, *Hazel Harrison* launches a full-time career as a pianist.

Her Story

PAGE № 87

1916

Continuing a lifelong pattern of activism supporting a woman's right to control her own body, *Margaret Sanger* opens the first birth control clinic. Sanger says: "Woman must have her freedom, the fundamental freedom of choosing whether or not she will be a mother and how many children she will have. Regardless of what man's attitude may be, that problem is hers–and before it can be his, it is hers alone. She goes through the vale of death alone, each time a babe is born. As it is the right neither of man nor the state to coerce her into this ordeal, so it is her right to decide whether she will endure it."

Painter *Georgia O'Keeffe* first exhibits some of her artistic works, especially her large flower canvases. By the end of her life, she completes more than two hundred flower paintings, in addition to other well-known works of art.

1916

Hetty Green, called the "Witch of Wall Street," dies; her estate of $100 million makes her the richest woman in America and, probably, the world.

1916

1916

Mary Pickford is the first movie star to form and own a film company and the first actor, male or female, to become a millionaire.

—1916—

—1917—

1917

1917

Committed pacifist *Jeannette Rankin* is the first woman elected to the U.S. House of Representatives. Rankin is one of a small number to vote against the United States entering World War I and, as a result, she is not reelected. Later, she returns to Congress and is the only member to vote against U.S. participation in World War II. Her commitment to pacifism culminates when, at the age of eighty-seven, she leads five thousand women on a march in Washington, D.C., protesting U.S. involvement in Vietnam. She said: "We're half the people; we should be half the Congress."

Dancer *Isadora Duncan*, considered the "Mother of Modern Dance," appears at the Metropolitan Opera House.

19-17

WOMAN YOUR COUNTRY NEEDS YOU!

STATE & NATIONAL COUNCILS of DEFENSE

SERVICE

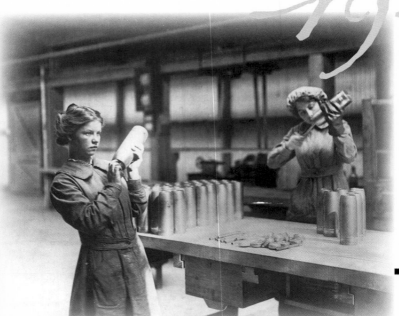

1917

1917

1918

The United States enters World War I. Employment opportunities for women expand significantly as so many men leave to fight.

1917

Microbiologist *Alice Evans* discovers the bacterium responsible for undulant fever; this leads to pasteurization of all milk.

Writer *Willa Cather* publishes the novel *My Antonia*, now considered one of her best works of fiction.

Singer and performer *Bessie Smith,* who becomes one of the greatest vaudeville blues singers, is discovered by "Ma" Rainey.

Maria Montoya Martinez is a Native American ceramic artist who creates distinctive black-on-black pottery.

1919

1919

19-18

19-19

1919

1918

Jessie Redmon Fauset is the editor of the NAACP's "Crisis" magazine.

1919

The research conducted by physician and pathologist *Louise Pearce* leads to a cure for African sleeping sickness.

Mabel Dodge Luhan moves to New Mexico and opens her home for "salons" of artists, writers, and photographers of note who live in the area.

Her Story

1919

Beginning around this year and continuing until the 1930s, the Harlem Renaissance is a cultural movement led by African Americans in New York City. Characterized by an outpouring of music, art, and literature, it addresses racial problems at the time and reflects black pride and optimism. Women play a significant role in this cultural growth. They include Jessie Redmon Fauset, Zora Neale Hurston, Ma Rainey, and Ethel Waters, among others.

1920

Charlotte Woodward Pierce is the only one of the attendees from the 1848 Seneca Falls Convention who lives to see women's suffrage enacted when the Nineteenth Amendment is ratified.

1920 *Mary Anderson* is the first director of the Women's Bureau of the U.S. Department of Labor.

1920

1920 In anticipation of the victory of suffrage, the National American Woman Suffrage Association organizes the nonpartisan League of Women Voters. A woman's right to vote becomes law on August 26, after seventy-two years of struggle.

1920

1920

Charleston, South Carolina, allows blacks to teach in the public schools as a result of a campaign in which *Septima Clark* (seated) and the NAACP participate.

Prolific writer *Edith Wharton* is the first woman to win the Pulitzer Prize in fiction, for her novel *The Age of Innocence*.

1921

1921

Betty Cook founds the Women's Bond Club of New York.

1921

M. Carey Thomas, who plays a large role in shaping Bryn Mawr College during her twenty-eight-year association with it, opens the college's Summer School for Women in Industry, to train women in union leadership skills. Thomas said: "Women while in college ought to have the broadest possible education. This college education should be the same as men's, not only because there is one best education, but because men's and women's effectiveness and happiness and the welfare of the generation to come after them will be vastly increased if their college education has given them the same intellectual training and the same scholarly and moral ideals."

Ellen Churchill Semple is the first woman elected president of the Association of American Geographers.

1921

World-renowned mathematician *Amalie Emmy Noether* writes a paper that lays the foundation for modern abstract algebra; she later teaches at Bryn Mawr.

1921

1921

Grace Abbott is the chief of the federal Children's Bureau, succeeding Julia Lathrop. She campaigns tirelessly to abolish child labor.

African American aviatrix and civil rights advocate *Bessie Coleman* is the first U.S. woman to win an international pilot's license.

Helen Wills Moody wins the National Junior Tennis Championship on her first attempt. Her outstanding athletic career in tennis continues through the 1920s and 1930s.

1921

1921

1921

Marjorie Child Husted invents the character Betty Crocker and uses this pseudonym for newspaper columns and cookbooks.

Comedian, actress, and singer *Fanny Brice* achieves stardom on Broadway when she sings the French torch song "My Man" in the Ziegfeld Follies. She is known for her character Baby Snooks and the song "Second Hand Rose," among others. Her story is told in the 1964 Broadway musical and 1968 movie *Funny Girl*.

1922

Lila Wallace cofounds the magazine *Reader's Digest*.

1922

1922

To get a property deed signed, President Harding appoints bilingual Seminole interpreter *Alice Davis* (seated second from left) first female chief of the Seminole people. When she insists on payment for the land, she is dismissed.

1922

Emily Post publishes the first edition of her etiquette book. By the time of her death, it has gone through ten editions and ninety printings.

1922

ETIQUETTE
PEGGY POST

Journalist, writer, and poet *Ida Husted Harper* completes a six-volume history of women's suffrage, bringing the story up to 1920.

1922

VOTES FOR WOMEN

1922

Attorney and suffragist *Florence Allen* is the first woman to win a judicial post.

1923

In *Adkins v. Children's Hospital*, the U.S. Supreme Court rules against a minimum-wage law for women and children in the District of Columbia.

1923

Entrepreneur *Ida Rosenthal* forms the Maidenform Brassiere Company.

1923

Ada Comstock Notestein is the first full-time president of Radcliffe College.

1923

1923

Poet *Edna St. Vincent Millay* is the first American woman to win a Pulitzer Prize in poetry for *The Ballad of the Harp-Weaver*. She is the most widely read poet of her generation. Her poem "First Fig" is well known:

My candle burns at both ends;
It will not last the night;
But ah, my foes, and oh, my friends–
It gives a lovely light!

1923

Actress *Maude Adams* starts a second career as a lighting designer. She develops an incandescent bulb widely used in color film projectors.

1923

After having debuted as a young teenager in a traveling minstrel show and performing for many years, *Ma Rainey* signs a recording contract with Paramount. The "Mother of the Blues" will record more than a hundred songs between 1923 and 1928.

Her Story

PAGE № 97

1924

Novelist *Edna Ferber* publishes *So Big* and wins a Pulitzer Prize.

1924

1924

Microbiologist and physician *Gladys Dick* codiscovers the microbe that causes scarlet fever and co-patents the Dick test (patented not to make money but to preserve its safety and purity) to determine an individual's susceptibility to scarlet fever, which claimed thousands of lives in epidemics.

1924

Writer, feminist, and political activist *Emily Newell Blair* works tirelessly for suffrage and then to organize women voters and train them as Democratic Party workers. In much of her writing and in her actions, she emphasizes the potential for political power among organized women.

Margaret Petherbridge Farrar is the first woman to produce a crossword puzzle book.

1925

Florence Sabin is the first woman elected to the National Academy of Sciences. In addition to training the next generation of researchers, she studies the role of the body's white cells in fending off diseases such as tuberculosis. She also makes important contributions to the histology of the brain. She is the first woman faculty member at the Johns Hopkins School of Medicine and its first full female professor.

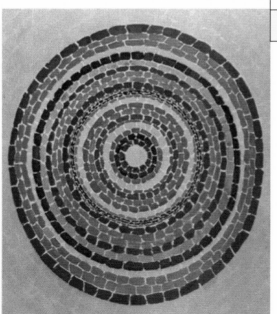

1924

Alma Thomas is the first graduate of Howard University's art department; she becomes a successful painter and teacher.

1925

1925

Novelist *Ellen Glasgow* achieves critical success as a writer; later she receives the Pulitzer Prize for *In This Our Life*.

Her Story

PAGE № 99

1925

Republican *Edith Nourse Rogers* is the first congresswoman from New England. She serves for thirty-five years and sponsors many significant bills, including those establishing the Women's Army Corps and the permanent nursing services of the Veterans Administration.

1925

Educator and artist *Polingaysi Qoyawayma* begins a thirty-year career teaching Hopi and Navajo students both their traditional school subjects and Native culture.

1925

Fashion designer *Hattie Carnegie* strikes a deal with I. Magnin that makes her clothing accessible to Hollywood's elite.

Versatile performer (comic, actress, singer, dancer) *Imogene Coca* makes her Broadway debut in *When You Smile*.

1925

Clark **GABLE**
William **POWELL** *Myrna* **LOY**

IN

MANHATTAN
MELODRAMA

PRODUCED BY
DAVID O. SELZNICK
DIRECTED BY
W. S. VAN DYKE

A
Cosmopolitan
Production

A
Metro-
Goldwyn-
Mayer
PICTURE

1925

Myrna Loy, a famous actress with more than sixty films to her credit, makes her screen debut in *Pretty Ladies*.

Josephine Baker is an African American dancer and entertainer. In 1925, she accepts an offer to appear in a Paris show, where she rises to stardom due to French enthusiasm for African American culture and jazz. She causes a sensation with her G-string ornamented with bananas.

1925

1925

1925

Mary Breckenridge founds the Kentucky Committee for Mothers and Babies, which becomes the Frontier Nursing Service.

1925

Pattie Field is the first woman in the U.S. consular service; she serves as the vice consul in Amsterdam.

CARL LAEMMLE presents
Carole LOMBARD in Faith Baldwin's
'LOVE BEFORE BREAKFAST'
with PRESTON FOSTER

1925

Well-known comedic actress Carole Lombard has her first starring role; she will appear in more than eighty feature-length films and short subjects. Her sense of humor is present both on the screen and off in jokes and pranks.

1926

Kate Smith lands her first Broadway role; later she is famous for singing "God Bless America."

Gloria Swanson is a famous silent screen actress who starts her own production company with Joseph Kennedy.

1927

1926

Alice Fong Yu is a teacher in the San Francisco Unified School District. As the only bilingual teacher for a predominantly Chinese-speaking student body, she wears many different hats, acting as translator, social worker, and all-purpose liaison.

–1926

1926

Gertrude Ederle is the first woman to swim the English Channel, breaking the previous record by nearly two hours.

1926

Violette Anderson is the first woman to receive a law degree in Illinois. She is also the first black woman admitted to practice before the U.S. Supreme Court.

1927

Jewish women start Women's American ORT (Organization for Rehabilitation Through Training). The organization, begun in 1880 in Russia, operates in every country where the Jewish population needs to learn a skill or trade.

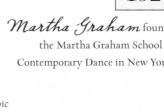

Martha Graham founds the Martha Graham School of Contemporary Dance in New York.

1927

Dorothy Parker is a witty and acerbic writer who joins the staff of the *New Yorker* magazine; she is one of the most successful and influential writers of her era.

1927

1927

Aviatrix *Ruth Elder* pushes for publicity for women in the field of flying. America takes notice of her and she is given the female lead in a silent movie.

1927

Mourning Dove is an Okanogan Indian who is a writer of Native American stories. She says: "Everything on the earth has a purpose, every disease an herb to cure it, and every person a mission."

Her Story

1927

Norwegian figure skater *Sonja Henie* wins her first world championship. She will win the next nine world amateur championships, as well as gold medals at the Olympics in 1928, 1932 and 1936; she becomes an American citizen in 1941.

TIME
THE WEEKLY NEWSMAGAZINE

SONJA HENIE
"Most always I win."
(Cinema)

1928

Anthropologist *Margaret Mead* publishes what will become her most famous book: *Coming of Age in Samoa*. She is also well remembered for her quote "Never doubt that a small group of thoughtful, committed citizens can change the world. Indeed, it is the only thing that ever has."

1927

1928

1928

Ruby Keeler is the first tap-dancing movie star.

1927

Singer *Ethel Waters* appears in the all-black musical revue *Africana* on Broadway. During her long career, she popularizes a number of hit songs, including "Stormy Weather" and "Dinah."

1928

Genevieve Cline is the first woman appointed a federal judge; her post is in New York at the U.S. Customs House.

Dog breeder *Dorothy Eustis* starts the Seeing Eye dog school to train dogs (primarily a special strain of German shepherds) to serve blind individuals.

Philanthropist and art collector *Marjorie Merriweather Post* provides funds for a Salvation Army food kitchen in New York and is known as "Lady Bountiful of Hell's Kitchen."

Agnes Moorehead (left) begins a long, successful acting career in radio, film, and television.

The onset of the Great Depression brings new support to the idea that women's place is in the home. Working women are said to be taking jobs away from men. Ironically, because of the poor economic situation, many women will find it necessary to work. **1929**

1929

Actress *Bette Davis* makes her Broadway debut in *Broken Dishes*.

Her Story

PAGE № 105

Ruth Seeger receives a Guggenheim Foundation Fellowship to study music and composing in Europe; she becomes an extraordinary classical composer and folk music activist.

1930

Two million women— a fifth of the female labor force—are office workers.

1930

1930

A singer and actress with a booming voice, *Ethel Merman* opens in *Girl Crazy* with the hit song "I've Got Rhythm."

Writer, pilot and navigator *Anne Morrow Lindbergh* is the first woman to gain a glider pilot's license. Later she is the first woman to receive the National Geographic Society's Hubbard Gold Medal for her part in the development of aviation routes. The theme of one of her inspirational books, *Gift from the Sea*, is the need for women to take the time to reflect and be alone.

1930

-1930

1930

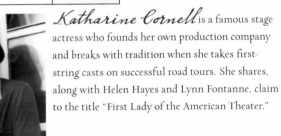

Katharine Cornell is a famous stage actress who founds her own production company and breaks with tradition when she takes first-string casts on successful road tours. She shares, along with Helen Hayes and Lynn Fontanne, claim to the title "First Lady of the American Theater."

1930

Historian *Edith Hamilton* publishes *The Greek Way*, a book that is both a critical and a popular success.

1930

Cherokee *Ruth Muskrat Bronson* is named the first guidance and placement officer of the Bureau of Indian Affairs.

1930

Screenwriter *Frances Marion* wins her first Oscar. She is the highest-paid screenwriter for over two decades.

1931

Lillian Gilbreth, cofounder of the field of industrial engineering, receives a medal from the Society of Industrial Engineers for her time studies. Referred to as the "First Lady of Engineering," she also is considered to be a pioneer in industrial and organizational psychology. She has twelve children, and their family experiences contribute to her theories of time and motion.

1931

1932

1930

Ellen Church (left), a trained nurse, becomes the first airline stewardess.

1932

On May 20-21, pilot *Amelia Earhart* is the first woman to fly solo across the Atlantic Ocean. She says: "Please know that I am aware of the hazards. I want to do it because I want to do it. Women must try to do things as men have tried. When they fail, their failure must be a challenge to others."

1932

Ruth Nichols is the first woman commercial airline pilot.

Her Story

Frances Perkins is appointed secretary of labor by President Franklin D. Roosevelt, becoming the first female cabinet member. She says: "I came to Washington to work for God, FDR, and the millions of forgotten, plain common workingmen."

1933

Journalist and reformer *Dorothy Day* starts the *Catholic Worker*, a monthly magazine that takes controversial positions on social issues.

1933

Ruth Patrick joins the Academy of Natural Sciences as a volunteer in microscopy, as the academy does not pay women scientists at this time. She founds the field of limnology and is a recognized expert on river pollution.

Actress *Ida Lupino* appears in her first of over sixty films; later she directs as well and is widely respected as a pioneer filmmaker.

1933

1933

First Lady *Eleanor Roosevelt* breaks with tradition by holding her own White House press conferences and allowing only women reporters. Throughout her long life, Roosevelt continues a public service career by courageously speaking and writing, both in the United States and abroad, about social issues and concerns. She is widely celebrated and respected. Among her many quotable comments, she says: "You gain strength, courage and confidence by every experience in which you stop to look fear in the face. . . . You must do the thing you think you cannot do."

1933

Katharine Hepburn is a movie star and a feminist on- and offscreen. She wins her first of four Best Actress Oscars in 1933 for her work in *Morning Glory.*

1933

Florence Beatrice Smith Price is the first black woman to have a symphony performed by a major American orchestra.

Her Story

PAGE № 109

1934

While a young teenager, singer *Ella Fitzgerald* wins a talent contest at the Apollo Theatre in Harlem. Known around the world as the "First Lady of Song," Fitzgerald wins twelve Grammy Awards and is the favorite female vocalist of many musicians, including most jazzmen.

1934

Ruth Benedict is an early American anthropologist; she publishes *Patterns of Culture*, which details the impact of human culture on personality.

1934

1934

Actress *Rosalind Russell* makes her screen debut.

Popular gospel singer *Mahalia Jackson* releases her first album. She will sing at Dr. Martin Luther King Jr.'s funeral in 1968.

1934

1934

Journalist and broadcaster *Mary Margaret McBride* premieres on radio using the name Martha Deane and a down-home southern drawl. Her daily program offers advice for women.

WELCOME THIS NEW DAY FOR WOMANHOOD

This summer you can experience a comfort and an assurance of daintiness you have never known before

SANITARY PROTECTION WORN INTERNALLY

1934

Early corporate pioneer and leader *Gertrude Tenderich* charters the Tampax Sales Corporation after buying the patent; she serves as president of the company.

1935

Muriel Rukeyser wins the Yale Series of Younger Poets Award at age twenty-one for her first book of poetry, *Theory of Flight.*

1935

Electrical engineer *Mabel MacFerran Rockwell* is the only woman actively involved in designing and installing the power-generating machinery at the Hoover Dam. She later works on increasing aircraft efficiency at Lockheed Aircraft.

1935

Member of a distinguished and prominent family, lawyer *Sadie Alexander* is the second black woman to earn a doctorate; she and her husband test the state of Pennsylvania's public accommodations law, which they had drafted.

1935

Mari Sandoz publishes *Old Jules* (after a previous rejection) and wins a $5,000 prize for the most interesting and distinctive work of nonfiction; the book is selected by the Book-of-the-Month Club.

-1935

LAURA INGALLS WILDER
Little House on the Prairie

ILLUSTRATED BY GARTH WILLIAMS

1935

Children's book writer and teacher *Laura Ingalls Wilder* publishes *Little House on the Prairie*, the best-known of a number of novels she writes detailing the pioneer life of early homesteaders. The books have been in print continuously since their publication and are considered classics of American children's literature.

1935

Businesswoman *Effa Manley,* a co-owner of the former Newark Eagles baseball team of the Negro League, runs the business end of the team for more than a decade. In 2006, she becomes the first woman elected to the Baseball Hall of Fame.

Her Story

PAGE № 111

Singer and jazz great *Billie Holiday* records four hits, including "What a Little Moonlight Can Do," that bring her wide recognition and launch her career as the leading jazz singer of her time.

1935

1935

Child star *Shirley Temple* wins a special Oscar from the Academy of Motion Picture Arts and Sciences; later, as Shirley Temple Black, she is active in civic affairs and is appointed a U.S. ambassador.

1935

1935

Dancer *Ginger Rogers* does everything that Fred Astaire does but backward and in high heels. Their fourth movie together, *Top Hat*, is released in 1935 and is a huge box office success, earning an Academy Award nomination for Best Picture.

1936

Photographer *Dorothea Lange* begins her travels through the South and the Dust Bowl, recording the effects of the Depression and drought.

Runner *Helen Stephens*, the "Fulton Flash," wins two gold medals at the 1936 Olympics. She sets a world record in the 100-meter event that stands for twenty-four years. Later, she is the first woman to create, own, and manage a semiprofessional basketball team, the Helen Stephens Olympic Co-Eds, which is active from 1938 until 1952.

1936

1936

Journalist *Dorothy Kilgallen* writes *Girl Around the World* based on her participation as the only woman in an around-the-world race; she then uses this as the basis for a film script. Later she becomes a television personality, appearing on quiz shows.

Margaret Mitchell publishes her novel of the noble antebellum South, *Gone with the Wind*. The following year, she wins a Pulitzer Prize for the book, and more than a million and a half copies are sold.

1936

1936

1936

Life magazine's first cover features a photo taken by prominent photojournalist *Margaret Bourke-White*.

1936

Entrepreneur and businesswoman *Margaret Rudkin* starts the Pepperidge Farm bakeries; it becomes a multibillion-dollar operation.

1936

Journalist and radio personality *Dorothy Thompson* begins the "On the Record" column at the *New York Herald Tribune*.

Her Story

1937

Rose Blumkin, affectionately known as "Mrs. B.," founds the Nebraska Furniture Mart, which features an entirely new selling concept: to buy in bulk and pass on the savings to the customer, generally at no more than 10 percent markup. Her motto "Sell cheap and tell the truth" leads the Mart to sixty-nine years of uninterrupted sales growth and to a handshake deal to sell her business to financier Warren Buffett in 1983 for $60 million.

1938

The work of *Marion Post Wolcott* at the Farm Security Administration leads to her reputation as a major social documentary photographer.

1937

1937

Margaret Wise Brown writes her first children's book, *When the Wind Blew;* later she writes the award-winning *Goodnight Moon.*

1937

Bacteriologist and distinguished pediatrician *Hattie Alexander,* through her research on infections and antiserums, develops the first effective treatment for the once-fatal influenzal meningitis.

A patent is issued to physicist *Katherine Blodgett* for nonreflecting glass. This glass is used today in cell phones, television screens, eyeglasses, cameras, microscopes, telescopes, and projector lenses.

1938

Marjorie Kinnan Rawlings authors the Pulitzer Prize–winning *The Yearling*, read by several generations of American children.

1938

Pearl S. Buck wins the Nobel Prize for literature for *The Good Earth*, her epic descriptions of peasant life in China, which was published in 1931 and received the Pulitzer Prize in 1932.

1938

1938

1938

In 1938 at the Palmer House in Chicago, a number of separate Republican women's clubs merge to create a national lobbying organization, which they call the National Federation of Republican Women's Clubs.

1938

Emma Tenayuca leads pecan workers in labor protests in San Antonio, Texas. Raised by her grandparents because of her parents' dire financial circumstances, Tenayuca, one of eleven children, joined the labor movement when she was sixteen. A lifelong activist and later a teacher, Tenayuca also spearheads the movement to secure a minimum-wage law.

1938

Catherine Stern, immigrates to the United States. She becomes a famous leader in the education of kindergarten children. She develops large colored rods (Stern rods), cubes, and colored blocks so children can visually understand math. Her work has had a lasting effect on elementary math education.

Her Story

PAGE № 115

1939

1939

Karen Horney publishes her book *New Ways in Psychoanalysis*, which contains a thorough critique of Freudian practice, including its views of women. After being disqualified as an instructor and training analyst at the New York Psychoanalytic Institute because of this publication, she founds the Association for the Advancement of Psychoanalysis and later the American Institute for Psychoanalysis.

1939

Prolific photographer *Berenice Abbott* is also an educator, inventor, author, and historian. She receives four patents for photographic devices. Over the span of five decades she creates a systematic documentary of New York in photographs, which she publishes as *Changing New York*.

1939

Thérèse Bonney is in Finland to photograph the preparations for the 1940 Olympics and becomes the only photojournalist to document the Russian invasion in November.

1939

Anthropologist *Elsie Clews Parsons* publishes her most important work, *Pueblo Indian Religion*. During her twenty-five-year career of field research and writing, she emerges as the leading authority on many tribes in North America, Mexico, and South America.

1939

At age eighteen, *Anna Lee Aldred* is the first woman jockey in the United States.

1939

The guitar and autoharp work of
Maybelle Carter (seated)
has a pivotal impact on the entire
direction of country music.

1939

1939

On April 9, singer *Marian Anderson* gives a concert at the Lincoln Memorial attended by seventy-five thousand people after the Daughters of the American Revolution refuse to allow her to sing at Constitution Hall because of her race.

1939

Zora Neale Hurston publishes *Moses, Man of the Mountain* and is firmly established as a major writer.

1939

Grandma Moses first
exhibits her paintings at the Museum
of Modern Art. Moses, who did not
take up oil painting until her late
seventies, when arthritis forced her to
give up embroidery, painted in what is
termed American primitive.

1940

Dale Messick is the first female cartoonist with her comic strip *Brenda Starr, Reporter*.

1940

May Arbuthnot co-authors the Dick and Jane series, which teaches millions of children how to read.

Hattie McDaniel, a movie actress, singer, and radio and television personality, is the first African American to win an Academy Award for her performance as Mammy in *Gone with the Wind*.

1940

1940

Judy Garland is an entertainer whose career spans many years. As a child star in the now famous film *The Wizard of Oz*, she receives a special Academy Award for her portrayal of Dorothy. Garland sings, dances, and acts in more than thirty films and another thirty television specials.

1940

Comedienne *Minnie Pearl* joins country music's Grand Ole Opry as its only female member.

1940 | Sculptor *Louise Nevelson* has her first one-woman show.

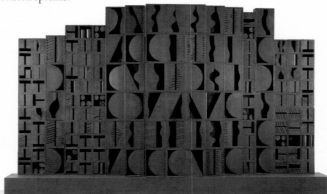

1940

Margaret Chase Smith is first elected as a U.S. representative from Maine. After serving four terms in the House, she wins election to the U.S. Senate in 1948, becoming the first woman elected to both houses of Congress. She is the first woman to have her name placed in nomination for the presidency by either major party. She serves four terms in the Senate and a total of thirty-two years in Congress.

On December 7, the Japanese bomb American planes and battleships in Pearl Harbor, Hawaii. This unprovoked attack, with its staggering death toll, causes the United States to enter World War II. The war will provide unprecedented employment opportunities for millions of American women.

1941

...we here highly resolve that these dead shall not have died in vain...

REMEMBER DEC. 7th!

No. 21 JAN.-FEB.
Wonder Woman
TEN CENTS
Wonder Woman and the ADVENTURE of the ATOM UNIVERSE!

1941

Fictional character *Wonder Woman* is introduced as a comic book heroine. With Superman and Batman, she is part of DC Comics' "Big Three." Wonder Woman achieves later popularity on television when Lynda Carter plays the role in the early 1970s.

1941

The Kashaya Pomo, a California Native American tribe, is guided by the spiritual leader *Essie Parrish* from 1941 to 1979. At age six, Parrish is acknowledged to be a "dreamer" or visionary of her people. She has the ability to both prophesy and interpret dreams. Parris is committed to developing a record of the Pomo people. She works with social scientists to make more than twenty anthropological films documenting Pomo culture and ceremonies, including a film of her life that wins an award at the 1969 Cannes Film Festival. She is also dedicated to educating tribal youth on their language, culture, customs, and laws.

1942

Jazz singer *Sarah Vaughan* wins an amateur contest at the Apollo Theatre singing "Body and Soul." Vaughan is one of the first female singers to incorporate bop phrasing into her singing. Over the course of her long career, she is repeatedly voted top female vocalist, and she is inducted into the Jazz Hall of Fame in 1990.

1941

1941

Cartoonist *Hilda Terry* draws the comic strip *It's a Girl's Life* (later known as *Teena*); it is first published by William Randolph Hearst, and will run in newspapers nationwide for twenty-three years. Terry becomes the first female member of the National Cartoonists Society in 1951.

The United States adopts as the standard the tuberculin test developed by *Florence Seibert.* The World Health Organization follows in 1942. Her test is still in use today.

1941

1942

Elizabeth Taylor makes her screen debut in *There's One Born Every Minute.*

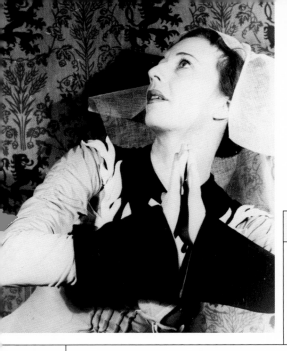

1942

Jackie Cochran creates the Women Airforce Service Pilots (WASP); later she is the first woman to break the sound barrier.

1942

Rodeo, the first major composition by choreographer *Agnes de Mille*, is danced by the Ballet Russe de Monte Carlo.

1943

Lena Horne, a well-known African American entertainer, stars and sings in *Stormy Weather*. Her rendition of the song by the same name becomes her trademark.

1942

1943

1942

Sylvia Porter changes the byline for her *New York Post* financial column from S. F. Porter to Sylvia F. Porter, as it is now acceptable for such a column to be written by a woman.

1942

Oveta Culp Hobby is the first director of the Women's Army Auxiliary Corps; later Colonel Hobby becomes the first secretary of health, education, and welfare.

Attorney *Constance Baker Motley* is a pioneer in both black civil rights and women's rights. She becomes the first black woman appointed to the federal judiciary.

1943

1943

Euphemia Lofton Haynes is the first black woman to receive a Ph.D. in mathematics.

Writer and lecturer *Ayn Rand* publishes her novel *The Fountainhead*, which is later made into a movie. This book and others she wrote espouse her unique philosophy, objectivism, which she characterized as a "philosophy for living on earth." Every book Ayn Rand writes during her lifetime is still in print, widely read and influencing later generations.

1943

1943

"Rosie the Riveter" becomes the often-used graphic representation of U.S. women who provide most of the labor to produce the materials necessary for World War II.

1943

Photojournalist *Georgette "Dicky" Chappelle* is one of the few women combat correspondents during World War II, the 1956 Hungarian revolution, the battles of Fidel Castro in Cuba, and the Vietnam conflict (where she was killed in 1965).

Activist, attorney, and ordained minister *Pauli Murray* overcomes both racial and gender discrimination and becomes the first black woman awarded a law doctorate from Yale. Later, she is the first black female priest ordained by the Episcopal Church.

1944

Pediatrician and cardiologist *Helen Brooke Taussig* develops a cardiac catherization operation that solves the often fatal condition known as "blue baby," saving countless infants.

1944

Actress *Angela Lansbury* receives her first Oscar nomination for her performance in *Gaslight*; her entertainment career spans more than sixty years.

1944

Nuclear physicist *Chien-Shiung Wu* works on the Manhattan Project, a secret effort to develop the atom bomb.

1943

Admiral *Grace Murray Hopper* is a professor of mathematics at Vassar College who is recruited to work on the first computer. Later, she creates the first computer compiler, the software that translates human language into the zeroes and ones that computers understand, paving the way for personal computers. She is also involved in the development of the business computer language COBOL.

1944

Her Story

1944

After World War II, in part to stagger the reentry of millions of people into the workforce, the GI Bill of Rights is passed. It provides for federal aid for education, low-interest loans for the purchase of homes and businesses, and hospitalization. These benefits substantially enrich American families and mean that higher education is no longer only available to the wealthy. Home ownership, for example, becomes more widely available, and many Americans move to new homes in suburban communities.

1945

Glamorous film actress *Joan Crawford* wins the Best Actress Oscar for her performance in *Mildred Pierce*. Over an acting career that spans almost fifty years, she plays many driven, independent women and is nominated for three Academy Awards.

1944

1945

1945

In 1945, the median annual income for all families and individuals is $2,379.

1944

Actress *Lauren Bacall* makes her film debut opposite Humphrey Bogart in *To Have and Have Not*.

1945

Entrepreneurs *Bertha Neaky* and *Ruth Bigelow,* (pictured) using old family recipes, start the Bigelow Tea Company. The name of one flavor, Constant Comment, is from a friend's remark "My dear, your tea was the hit of the party; there was nothing but constant comment."

The Links, Inc., is established as a social organization to encourage black women to gain a richer appreciation of the arts and culture and to become active in their communities.

1946

1946

Historian and feminist *Mary Ritter Beard* studies women in American history. She writes several books and lectures on women's positions in the labor movement and as activists for social change.

Most of the more than six million American women who entered the workforce during World War II are now, after the war's end, being pushed out of traditionally male jobs.

1945

1946

1946

Sociologist, political scientist, and pacifist *Emily Greene Balch,* who is a leader of the women's movement for peace, is awarded the Nobel Peace Prize.

1945 The first twelve women enter Harvard Medical School.

Her Story

1946

Actress *Jayne Meadows* appears in *Lady in the Lake*. A critically acclaimed dramatic and comedic actress, she is an Emmy Award winner and five-time Emmy nominee. Later in her career, she receives the Susan B. Anthony Award for her contribution in portraying women in positive roles. She also tours the United States for six years in a one-woman show, *Powerful Women in History*.

1946

1946

Estée Lauder starts her cosmetics empire by selling face creams that she both formulates and makes herself. She said that she relied on three means of communication to build her multimillion-dollar empire: "telephone, telegraph, and tell a woman."

1946

1946

Edith Houghton, who plays baseball so well that at one time or another she plays every position, is the first woman scout for a major league baseball team when the Philadelphia Phillies hire her.

1946

Actress and cultural icon *Marilyn Monroe*, who will become famous more for her beauty and breathy voice than for her acting talent, signs a short-term contract with 20th Century Fox. Her appearance in the 1950 movie *All About Eve* leads to another contract with Fox, and she becomes a household name after a large-scale publicity campaign.

Businesswoman *Dorothy Stimson Bullitt* founds KING Broadcasting, based in Seattle. It is one of the nation's first privately owned broadcast empires.

1947

1947

Actress *Jessica Tandy* appears on Broadway in *A Streetcar Named Desire*, for which she wins a Tony Award. Her long and distinguished career culminates in a Best Actress Oscar in 1989 (at eighty, she is the oldest actor to receive a nonhonorary Oscar) for *Driving Miss Daisy*.

1947

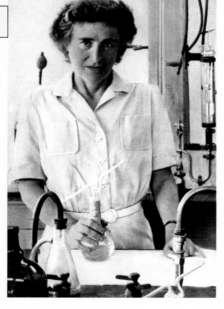

Biochemist and physician *Gerty Radnitz Cori* is the first American woman to win the Nobel Prize in the sciences. Her research involved studying the overall process of carbohydrate metabolism in the body, later called the Cori cycle.

1947

1947

Alice Hamilton is the first woman to receive the Lasker Award of the U.S. Public Health Association. Dr. Hamilton is considered to be the founder of the field of occupational medicine.

1947

Marjory Stoneman Douglas is considered the "patron saint of the Everglades"; she writes *The Everglades: River of Grass* and raises awareness of the need for environmental conservation efforts.

Her Story

PAGE № 127

1947

Cuban-born singer *Celia Cruz* records her first songs in Venezuela; she is considered the undisputed queen of salsa music. Posthumously, she will be awarded the Congressional Gold Medal; she is the first Hispanic woman to receive it.

1948

Mary Agnes Hallaren is the first nonmedical woman to be a regular army officer. At this time, women serve in a separate unit, the Women's Army Corps. Hallaren is part of the effort to combine military men and women into one organization.

1947

1948

1948

High jumper *Alice Coachman* becomes the first black woman to win an Olympic gold medal.

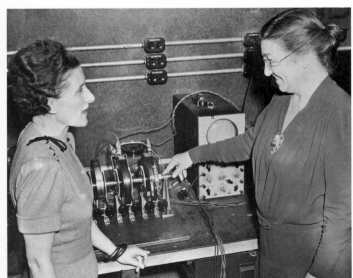

1947

Edith Clarke (right) is one of the first three women fellows of the American Institute of Electrical Engineers. During her accomplished career, she publishes what becomes the standard textbook for circuit analysis calculations and patents a method for regulating transmission line voltage.

1949

Stella Adler

teaches the principles of
acting, character, and script
analysis and is the leading
American proponent of
Method acting.

1949

Yoshiko Uchida

publishes the first of her
twenty-eight books for
children: *The Dancing Kettle
and Other Japanese Folk Tales.*

1949

1949

While suburbs existed before World War II, after the
war a massive population shift begins to suburbs that
are dependent on automobiles for transportation.
This deeply affects the lives of many women
(primarily middle-class whites) who stay home to
raise their children and spend much of their time
chauffeuring them from activity to activity.

1949

Gwendolyn Brooks

publishes a collection of verse,
Annie Allen, for which she will
become the first black poet
to win the Pulitzer Prize.

1949

Entertainer *Mary Martin* stars in *South Pacific,* which opens this year. During her career she plays the lead in a number of famous musicals.

1949

1949

Georgia Neese Clark is the first woman to be treasurer of the United States.

1949

Romana Banuelos starts a tortilla factory, which becomes Ramona's Mexican Food Products. Later she serves as treasurer of the United States from 1971 to 1974.

1949

Actor, writer, and producer *Gertrude Berg* transfers her radio success to television with *The Goldbergs,* the first family sitcom.

1950 *Malvina Reynolds* is a songwriter and performer who begins as a folk music artist at age fifty, after a career as a newspaperwoman.

1950

Rachel Fuller Brown (right) and *Elizabeth Lee Hazen* (left) develop the first antifungal antibiotic, which they name nystatin.

1950

1950

1950

Classical soprano *Roberta Peters* makes her Metropolitan Opera debut; subsequently she performs at the White House for every president since John F. Kennedy.

1950

Cofounder of Beech Aircraft, *Olive Ann Beech* becomes president of the corporation. Her career success led to her being dubbed the "First Lady of Aviation."

Her Story

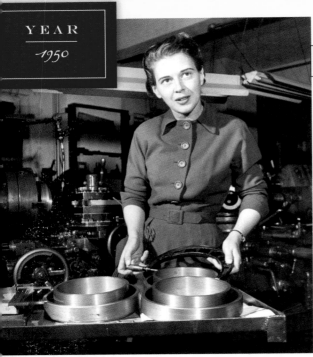

1950

Electrical engineer *Beatrice Hicks* helps found the Society of Women Engineers at a time when less than 1 percent of U.S. engineers are women.

1950

Jade Snow Wong is a Chinese American writer whose novels depict the Chinese culture in the United States; she is also a well-recognized pottery artist. Her autobiographical novel, *Fifth Chinese Daughter*, is published in 1950 and selected for the Book-of-the-Month Club.

FIFTH CHINESE DAUGHTER

JADE SNOW WONG

ILLUSTRATED BY—KATHRYN UHL

1950

Margaret Donahue is the first woman executive in major league baseball.

1950

Babe Didrikson Zaharias is named woman athlete of the half century; she is an all-around athlete who dominates every sport she tries.

1950

Abstract Expressionist painter *Helen Frankenthaler* uses a unique technique called "staining," in which thinned-down oils soak through unprimed canvas.

1950

Althea Gibson becomes the first black woman to play tennis in the U.S. Open. She breaks tennis's color barrier at Wimbledon and wins. Subsequently, she is given a parade in New York City. Gibson says: "I always wanted to be somebody. If I made it, it's half because I was game enough to take a lot of punishment along the way and half because there were a lot of people who cared enough to help me."

1950

1950

Writer *Hisaye Yamamoto* (far right) publishes her story "The Legend of Miss Sasagawara," drawing upon her experiences of having been among the Japanese Americans interned in detention camps in the United States during World War II.

1951

Abstract Expressionist painter *Lee Krasner* debuts her artwork. She later moves into a large geometric abstract style.

1951

Talented actress and businesswoman *Lucille Ball* launches her long-running, award-winning comedy television series *I Love Lucy*.

-1951-

1951

Outspoken, versatile actress *Shelley Winters* receives her first Oscar nomination, for her work in *A Place in the Sun*; she will go on to receive three more nominations and win two Academy Awards during her long film career, spanning six decades. She never forgets her Jewish heritage, donating her Oscar (1959) for her role in *The Diary of Anne Frank* to the Anne Frank Museum in Amsterdam.

Creative salesperson and entrepreneur *Brownie Wise* is named vice president of Tupperware; she encourages thousands of homemakers to become businesswomen through the concept of home sales parties.

1951

Maggie Higgins is the first woman to win the Pulitzer Prize for international reporting. She is the only woman reporter during the Korean conflict, where she reports from the battlefront for twenty-three months.

German-born philosopher *Hannah Arendt* is a controversial author who argues complex social theory. Her first major work, *The Origins of Totalitarianism* (written after she became a naturalized U.S. citizen), is a response to the devastating effects of Stalinism and Nazism. She is the first woman to hold the rank of full professor at Princeton University.

1951

1951

Lillian Vernon is one of the pioneers of direct mail marketing. The Lillian Vernon Corporation becomes so profitable that she is the first woman to have a company listed on the American Stock Exchange.

1951

Amateur artist *Bette Nesmith Graham* (front far right) works in her kitchen to make and market Mistake Out, a white liquid that covers up errors in typed documents. This product is eventually renamed Liquid Paper; she later sells the product to Gillette for $47 million.

Her Story

1952

Flutist *Doriot Anthony Dwyer* (a descendant of Susan B. Anthony) is the first woman appointed to a principal chair in a major orchestra.

1952

1952

Leontyne Price makes her professional debut as an opera singer.

1952

Lillian Hellman, a famous playwright, testifies before the House Un-American Activities Committee. She is blacklisted, which means she is prevented from finding paid employment, but she continues her social activism. In the early 1960s, she finds work again and begins her memoirs by publishing *An Unfinished Woman;* it later receives the National Book Award.

Southern writer *Flannery O'Connor* publishes her first novel, *Wise Blood*; she is best known as a master of short stories.

1952

Educator *Frances Horwich* develops *Ding Dong School*, the first TV program to be aimed at a preschool audience and to provide a quality learning experience.

Barbara Holdridge (on left) cofounds Caedmon Records with her friend *Marianne Mantell*. They begin their career of recording spoken voices of famous poets and writers by signing their request letters with initials rather than their names, so their subjects would not know they were women. Today, audiobooks are a $2 billion industry.

1952

Athlete *Gertrude Dunn* is voted Rookie of the Year in the All American Girls Professional Baseball League. She also excels in field hockey and lacrosse.

1952

Her Story

PAGE № 137

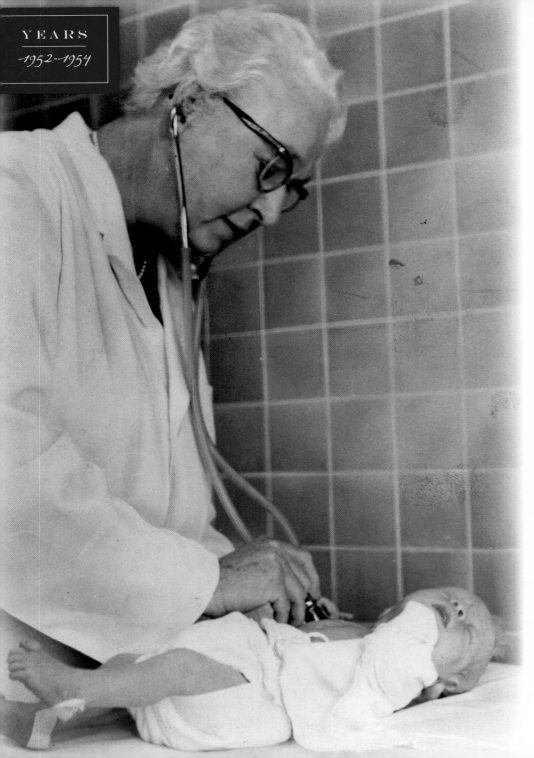

1952

1952

Virginia Apgar is a physician who develops a series of rapid checks for use on newborn infants to determine if the babies need medical attention. The Apgar numerical score, determined at one minute and again at five minutes after birth, has been used worldwide ever since.

1952

1953

Mary Steichen Calderone becomes the medical director of Planned Parenthood; she organizes a conference for women health providers on the dangers of illegal abortion.

1954

Native American dancer *Maria Tallchief* becomes the prima ballerina with the New York City Ballet.

1953

Politician and diplomat *Clare Boothe Luce* is appointed the U.S. ambassador to Italy.

1953

1953

Lupe Serrano joins the American Ballet Theatre as the first female Hispanic principal dancer, having made her professional career debut at age thirteen in Mexico, where she became a leading ballerina. By the time of her retirement in 1971, she will have been principal ballerina and danced more than fifty roles.

1953

Comedic actress and animal rights advocate *Betty White* stars in her first sitcom, *Life with Elizabeth*.

Her Story

PAGE № 139

1955

Edith Green begins a two-decade-long career in Congress as a representative from Oregon. She works on women's rights and plays a significant role in securing the passage of Title IX, the first comprehensive federal law to prohibit sex discrimination against students and employees of educational institutions. Title IX benefits both males and females but is widely recognized as helping to develop parity for women in athletics.

1955

Popular singer and radio and television personality *Dinah Shore* wins her first Emmy as host of her TV variety show. During her long career, she is exceptional in her ability to connect with her audiences.

Shirley MacLaine appears in her first film, the Alfred Hitchcock classic *The Trouble with Harry.* Over the course of her career she appears in more than fifty films and many TV specials and stage shows, and writes numerous books about her spiritual beliefs.

1955

1955

Opera singer *Beverly Sills* joins the New York City Opera. During her long singing career, she records eighteen full-length operas. After she retires, she devotes herself to charity work for the prevention of birth defects.

1955

1955

The arrest of *Rosa Parks* for refusing to give up her seat on a bus to a white man sparks the Montgomery, Alabama, bus boycott. Black women, the system's main users, support the boycott for more than a year. Parks' courageous action is seen as an underpinning of the civil rights movement. In 1996 she is awarded the Presidential Medal of Freedom, and in 1999 she receives the Congressional Gold Medal. Of the events of the fateful day, she said: "The only tired I was, was tired of giving in."

Her Story

PAGE № 141

1956

Financier *Josephine Perfect Bay* is the president of A. M. Kidder and Company and the first woman to head a member firm of the New York Stock Exchange.

1956

Margaret Towner is ordained as the first female Presbyterian minister. She says: "Find where your talents are best used and where you would feel comfortable being involved. Help to build a world where there is peace. Work to be healers."

1956

Seven-time Grammy Award–winning singer *Tina Turner* begins her career when she is discovered by Ike Turner, with whom she subsequently suffers a turbulent marriage; she works with him until the mid-1970s and then goes on to have a successful solo career.

1956

The United Methodist Church grants full ordination and clergy rights to women.

1956

1956

Sociologist, educator, and writer *Rose Hum Lee* produces a pioneering study of Chinese American communities in the United States. She is the first Chinese American woman to chair a university department.

La Leche League is founded; its mission is to help mothers to breast-feed by supporting a better understanding of the benefits of breast feeding for both mother and baby.

1956

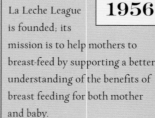

1957

Social justice activist *Ella Baker,* who was instrumental in the founding of the Student Non-Violent Coordinating Committee (SNCC), is one of the founders, with Martin Luther King Jr., of the Southern Christian Leadership Conference.

1957

Educator *Sadie Ginsberg* cofounds the Children's Guild; later she and Elinor Guggenheim start the Day Care and Child Development Council to push for government-backed day care and early childhood education.

1957

Alice Herrington founds the nonprofit organization Friends of Animals and, in 1967, the Committee for Humane Legislation. Friends of Animals is the first national organization to promote low-cost neutering of dogs and cats; they work to ensure cruelty-free treatment of all animals.

1956

1957

Dominican nun *Sister Rose Thering* is instrumental in having the Vatican rescind the view that the Jews killed Christ. For over forty years, through her commitment and advocacy, she is instrumental in creating programs to educate people on the evils of the Holocaust.

1957

Chita Rivera is an actress, singer, and dancer who electrifies Broadway audiences with her performance as Anita in *West Side Story*.

1957

1957

Country-western singer and songwriter *Patsy Cline* gets a big break singing "Walking After Midnight" on Arthur Godfrey's *Talent Scouts* on television. She dies at age thirty, but her songs are still heard in recordings done by other artists and in her own recorded works.

Her Story

Chef *Joyce Chen* opens a Chinese restaurant in Cambridge, Massachusetts. She begins to write cookbooks and hosts a cooking show on public television. She is credited with popularizing Mandarin Chinese food in the United States.

1958

1959

1958

Educator *Ethel Percy Andrus* starts the American Association of Retired Persons (AARP) to create a national lobbying voice and activist group that addresses the concern of older Americans.

Ruth Handler, co-owner of Mattel Toys, Inc., introduces the Barbie doll. Since its inception, more than one billion have been sold. One of Handler's novel ideas was to sell different clothing and accessories for the doll, which contributed to the product's enduring success.

1958

1959

1959

A Raisin in the Sun, by *Lorraine Hansberry,* makes its Broadway debut. It is the story of the working-class Younger family, who wish to leave behind their tenement apartment, and their struggle to find their place in life amidst life's trials. It has an entirely African American cast and is the first Broadway play to be written by a black woman. Nominated for four Tony Awards, it is made into a movie in 1961.

1959

Entertainer *Carol Burnett* has her Broadway debut in *Once upon a Mattress.* Her comedic timing is a hallmark of her performance style both on television and in plays.

1960

Folk singer and songwriter *Joan Baez* releases the first of her six gold albums. Throughout her singing career, she continues as a strong political activist. She tours on behalf of such causes as African famine relief and Amnesty International.

Pioneering aviatrix *Jerrie Cobb* is the first woman military pilot to pass all of the qualification tests administered to male astronaut candidates.

1960

1960

Appointed as a magistrate in Alaska, educator *Sadie Neakok* helps the Inupiat Eskimo people rightfully benefit from the U.S. legal system. The "Mother of Barrow" is a recognized humanitarian who helps all in need.

1960

1960

1960

Suzanne Farrell wins a scholarship to the School of American Ballet, handpicked by George Balanchine. Balanchine teaches her throughout her long career. She dances lead roles for three decades and goes on to be a prominent teacher of dance.

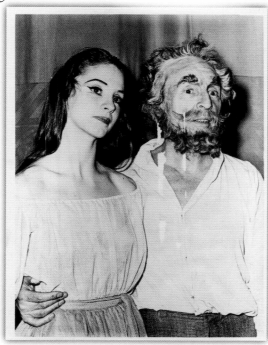

1960

Harper Lee writes *To Kill a Mockingbird*, which wins the Pulitzer Prize and is adapted for the stage and also made into a film.

TO KILL A
Mockingbird

A NOVEL BY
HARPER LEE

Her Story

PAGE № 145

At the Rome Olympics, runner *Wilma Rudolph* becomes the "fastest woman in the world" and the first American woman to win three gold medals at a single Olympics. Rudolph, who was the twentieth of twenty-two children, suffered from polio as a child and was told she would not be able to walk. But with physical therapy, the aid of her family, and strong determination, she becomes an outstanding athlete.

Women's roles change dramatically during the 1960s; they are in the mainstream, in roles formerly male-only, as primary economic providers and in positions of public authority. Women constitute 38 percent of the workforce in 1960. They are doctors, lawyers, engineers, scientists, construction workers, and firefighters, among other professions.

1960

The first birth control pill is approved for sale in the United States.

1960

1960

1960　　*1964*

1960

Mathematician *Irmgard Flugge-Lotz* becomes the first woman full professor at Stanford University. She had previously been denied a tenure-track position, being relegated to serving only as a lecturer, due to nepotism rules (prohibition on hiring scholars who are related to a current faculty member). She is recognized internationally for her contributions to aerodynamics and automatic theory control.

1960

Broadcast journalist *Nancy Dickerson Whitehead* is named by CBS News as the first woman correspondent to cover the White House.

1961

Writer and urban theorist *Jane Jacobs* publishes her classic book *The Death and Life of Great American Cities*, which urges a revolutionary rethinking of the ways in which neighborhoods and urban development are conceived.

1961

Eunice Kennedy Shriver helps establish a presidential committee on mental retardation.

1961

In 1956, at the age of forty-four, writer *Tillie Olsen* publishes *Tell Me a Riddle*, a collection of short stories, the first of which wins the O. Henry Award.

1961

Activist *Esther Peterson* influences President John F. Kennedy to establish the President's Commission on the Status of Women. At that time, she is also head of the federal government's Women's Bureau. Her long career of public service spans more than fifty years, during which she is an advocate for women's rights, labor concerns, and consumer issues such as truth in advertising and food purity.

Her Story

PAGE Nº 147

1962

Writer and organizer *Felice Schwartz* establishes Catalyst, an organization that works with corporations to promote women's leadership and to ensure more top-level opportunities for women.

1962

Historian *Barbara Tuchman* publishes *The Guns of August*, an account of the first month of World War I, for which she wins her first Pulitzer Prize. she wins a second Pulitzer Prize in 1972 for *Stilwell and the American Experience in China, 1911–1945*.

1962

1962

Novelist and short story writer *Katherine-Anne Porter* publishes *Ship of Fools*.

1962

Biologist *Rachel Carson* publishes *Silent Spring*, an exposé on the dangers of pesticides, including DDT.

1962

Entertainer *Barbra Streisand* makes her first appearance on Broadway; with forty-seven gold albums over the course of her career, she is in second place on the all-time list, trailing only Elvis Presley. She is also a film actress and director.

1962

1962

Labor leader and social activist *Dolores Huerta* helps found the United Farm Workers (UFW); she also helps organize the successful boycott of California's table grapes. The boycott lasts five years but results in the entire California table grape industry signing a three-year collective bargaining agreement with the UFW. She exhorted others to participate by saying: "Walk the street with us into history. Get off the sidewalk."

1962

Pharmacologist and physician *Frances Kelsey* prevents the drug thalidomide, which causes birth defects, from entering the United States.

1962

Performer *Rita Moreno* wins an Academy Award for her portrayal of Anita in *West Side Story*; she is one of a select few to win an Oscar, a Tony, a Grammy, and an Emmy as well as the Presidential Medal of Freedom.

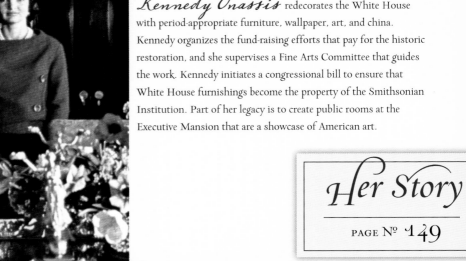

1962

A major project undertaken by First Lady *Jacqueline Kennedy Onassis* redecorates the White House with period-appropriate furniture, wallpaper, art, and china. Kennedy organizes the fund-raising efforts that pay for the historic restoration, and she supervises a Fine Arts Committee that guides the work. Kennedy initiates a congressional bill to ensure that White House furnishings become the property of the Smithsonian Institution. Part of her legacy is to create public rooms at the Executive Mansion that are a showcase of American art.

Her Story

1963

Jean Nidetch plans, develops, and incorporates Weight Watchers International. She says: "It's choice–not chance–that determines your destiny."

1963

Maria Goeppert-Mayer is the first U.S. woman to win the Nobel Prize in physics, for her work relating to the structure of atoms.

1963

Entrepreneur *Mary Kay Ash* founds Mary Kay Cosmetics; the company trains thousands of women who lack other career options to pursue direct sales careers and achieve financial success. The company culture provides an impetus to enthuse and motivate the employees. This multibillion-dollar company, now international in scope, established an associated charitable foundation in 1996. The foundation's mission is to cure cancers that particularly affect women and to end violence against women.

1963

1963

Navajo *Annie Dodge Wauneka* is the first Native American to receive the Presidential Medal of Freedom. She is recognized for working to eradicate tuberculosis and to reduce infant mortality. Wauneka said: "Over the years, I have learned that one failure, or even half a dozen failures should never be the end of trying. I must always try and try again, and I will continue to try as long as there is breath to do so."

1963

Encouraged by her mother to write and to make something of her life, author and feminist *Betty Friedan* publishes the highly influential *The Feminine Mystique*, a well-researched portrait of the gender roles of the time. The book examines and questions how women define themselves. It is often regarded as the impetus for the resurgence of the women's movement.

Her Story

1963

Julia Child starts her television cooking show, *The French Chef,* following the 1961 publication of her book *Mastering the Art of French Cooking.* Through her many cookbooks and her television series, she introduces French cooking to Americans and inspires a generation of chefs. She says, "Find something you're passionate about and keep tremendously interested in it."

1963

The Presidential Commission on the Status of Women issues its report recommending equal pay for comparable work; this leads to the Equal Pay Act of 1963, signed by President Kennedy.

1963

Eight-year-old *Nancy Lotsey* is the first girl admitted to the New Jersey Small-Fry Baseball League.

1963

Visual artist *Elaine de Kooning* finishes paintings of President John F. Kennedy for the Truman Library.

1964

Patricia Neal receives an Oscar for her performance in *Hud*. In 1965, at the height of her acting career (and only thirty-nine years old), she suffers three strokes in one evening. Her recovery is remarkable. She goes on to write about her ordeal, highlighting the need for persistence and determination in the rehabilitation of stroke victims. Worldwide, she is a symbol of both hope and recovery. She says: "A strong positive mental attitude will create more miracles than any wonder drug."

Avant-garde artist, musician, composer, and filmmaker *Yoko Ono* opens a conceptual and performance art event. In 2004, a forty-year retrospective of Ono's work receives high accolades and the International Association of Art Critics USA Award for Best Museum Show Originating in New York City.

1964

1963   *1964*

1964

Dynamic speaker, civil rights activist, and organizer *Fannie Lou Hamer* makes an unauthorized appearance at the 1964 Democratic National Convention to request inclusion for African Americans. She plays a significant role in voter registration in the South and is remembered for her daily recitation: "I am sick and tired of being sick and tired." She said about her work: "You know I work for the liberation of all people, because when I liberate myself, I'm liberating other people."

1964

The murder of *Kitty Genovese* in Queens, New York, is witnessed by almost forty neighbors, who ignore her cries for help. This provokes national discussion of urban isolation and indifference.

Her Story

PAGE № 153

1964

Therapist *Virginia Satir* publishes a book on family therapy that will become a classic in the field.

One of the world's foremost ceramic sculptors, potter *Ruth Duckworth,* a German-born Nazi refugee, comes to the United States from England to teach at the University of Chicago. Duckworth creates an immense and important body of work and is significant in continuing the modernist tradition.

1964

1964

Entertainer *Carol Channing* wins a Tony Award for her portrayal of Dolly Gallagher Levi in *Hello Dolly!*

1964   The Civil Rights Act of 1964 is landmark federal legislation. Segregation and discrimination, which have been tolerated for decades, are to be prohibited in public facilities, government, and employment. While the bill is being debated in Congress, discrimination based on sex is added at the last minute. Despite an extensive Senate filibuster, the bill passes. Women are thus explicitly included in this important legislative shift in U.S. public policy.

1964

Choreographer *Twyla Tharp* has her first piece publicly performed at Hunter College; she goes on to a long choreography career and forms her own dance company. She says: "I thought I had to make an impact on history. I had to become the greatest choreographer of my time. That was my mission. Posterity deals with us however it sees fit. But I gave it 20 years of my best shot."

World-renowned scientist *Judith Graham Pool* identifies Factor VIII, the clotting factor in human blood plasma, and develops a way to manufacture it. Hemophiliac patients can inject themselves with this factor, allowing them to live more fully.

Chemist *Stephanie Kwolek* creates a synthetic fiber, Kevlar, that is stronger than steel and is used in a variety of products, from bulletproof vests to space vehicles.

1965

Lady Bird Johnson, the most politically active First Lady since Eleanor Roosevelt, works to beautify America by helping ensure passage of the Highway Beautification Act of 1965. The act limits the number of billboards on major roadways. Johnson becomes well known for encouraging planting of wildflowers.

Racer *Shirley Muldowney* breaks the gender barrier by becoming the first woman licensed to compete in the supercharged gasoline dragster category by the National Hot Rod Association. Later, she becomes the first woman licensed in the top fuel category. Her many wins help open the doors for other women to compete in motor sports.

Ruth Fertel is the founder of Ruth's Chris Steak House in New Orleans, Louisiana. She pioneers a new idea to create a fine-dining chain. Her chain grows to become one of the largest and most successful restaurant operations in the world.

Her Story

PAGE № 155

-1966

1966

The National Organization for Women (NOW) is founded by Betty Friedan and other delegates to the third National Conference of Commissions on the Status of Women held in Washington, D.C., a significant step in the reemergence of the feminist movement in the United States. The goal of NOW is to bring about equality for all women, eliminate discrimination, secure reproductive rights, and end all forms of violence against women.

Yvonne Brathwaite-Burke begins her political life with her election as the first black assemblywoman in California; her long public service career includes offices at the national, state, and local government levels.

1966

1966

Philosopher and author of seventeen books *Susan Sontag* publishes essays entitled *Against Interpretation*.

1966

The television series *That Girl* premieres, featuring actress *Marlo Thomas* as a young, independent career woman. Later, Thomas develops a record album, books, and television specials that help children explore the message "You can achieve anything you want." During Thomas' acting career, she receives many prestigious awards. She uses her fame and her money to support social causes and continues the tradition of her late father, Danny Thomas, as the public face for St. Jude's Children's Hospital for Cancer Research.

1966

Opera conductor *Eve Queler* conducts her first public performance; as the music director for the Opera Orchestra of New York for more than thirty-four years, she conducts more than ninety operas.

Physicist *Betsy Ancker-Johnson* develops and patents a high-frequency signal generator. Later she is appointed by the president to be assistant secretary for science and technology.

1966

Brigadier general *Wilma Vaught* is the first woman to deploy with a Strategic Air Command (SAC) bombardment wing. Later, she is also the driving force in establishing the Women in Military Service for America Memorial Foundation, which raises the money to build the Women's Memorial in Arlington National Cemetery, honoring the nearly two million women who have served.

1966

The federal courts force the state of Texas to draw new legislative districts to end the gerrymandering that denies black people the opportunity to hold public office. Brilliant attorney and charismatic speaker *Barbara Jordan* is elected to the Texas state senate, the first black person since 1883 and the first black woman to hold the position.

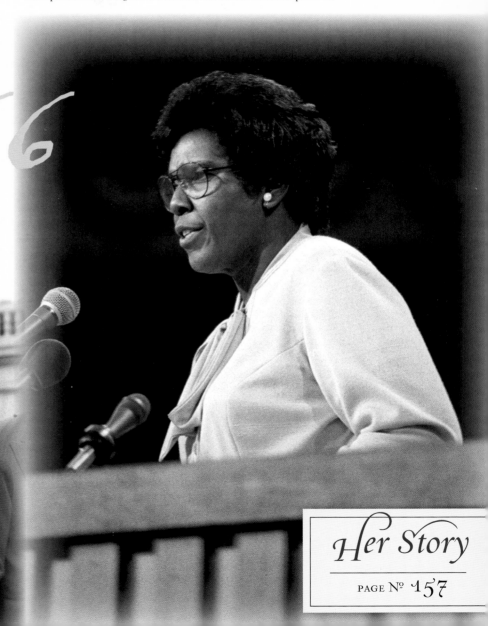

Her Story

1967

Muriel Siebert is the first woman to own a seat on the New York Stock Exchange (NYSE); today Muriel Siebert and Co. is the only woman-owned NYSE brokerage firm with a national presence. Siebert says: "Know the more that you succeed, the more you will be challenged. Many think that the biggest battle I had to fight was to buy my seat on the New York Stock Exchange. Yes, that was a battle, but that only got me into the game."

1967

Activist *Mary Sinclair* starts a long campaign of protest against the construction of nuclear power plants. Later, she also helps to develop creative solutions to environmental waste when she joins a number of others with diverse points of view, including Dow Chemical, through the National Resources Defense Council.

1967

K. Switzer secretly enters the Boston Marathon. When an official realizes that "K." stands for Kathrine, he tries to tear off her number, but she avoids him and finishes the race. She says: "I decided to use this experience to insure that other women who wanted to run would not be subjected to the same treatment. I became an organizer and an outspoken proponent for women's physical capability. . . . I felt the most important thing I could do for women was to create the forum for their acceptance in sports."

1967

Activist *Helen Claytor* becomes the first black president of the national YWCA.

1967

1967

Ida Rolf is recognized as the founder of the field of structural integration, a massage and realignment methodology that teaches trained therapists to reshape the body's connective tissue, resulting in better health and overall energy.

1967

1967

President Lyndon Baines Johnson signs Executive Order 11375, extending Executive Order 11246 to include gender as a protected category. This now requires that affirmative action be taken on behalf of women as well as minorities so that hiring is in line with gender proportions as well as racial proportions in the relevant labor pools.

1967

Primatologist *Dian Fossey* goes to Rwanda to study mountain gorillas. She spends most of her professional career with the gorilla families she studies.

1968

Writer *Erma Bombeck* looks to familiar household situations and points out their humorous aspects. Her first book, *At Wit's End*, is published. Later, her columns appear in seven hundred newspapers across the country.

1968

1968

Actress *Diahann Carroll* breaks television's color barrier by having a television show (*Julia*) built around her as the central character. She portrays a nonstereotypical character and wins a Golden Globe Award.

1968

Mary Washington Wylie founds Washington, Pittman and McKeever, which is one of the largest black-owned accounting firms in the country. She is the first African American woman to become a certified public accountant (CPA).

1968

Author *Joan Didion* receives national recognition when she publishes *Slouching Toward Bethlehem*, a collection of essays.

1968

Comedic actress *Goldie Hawn* joins the television program *Rowan and Martin's Laugh-In*; later she wins an Academy Award for her work in *Cactus Flower*. After demonstrating that women can pursue independent, fruitful lives in the movie *Private Benjamin*, Hawn becomes a movie producer.

Figure skater *Peggy Fleming* wins America's only Olympic gold medal at the Grenoble games.

1968

1968 Close to four hundred feminist protestors demonstrate against the Miss America Pageant because they believe it supports racist and sexist messages.

1968

1968

Shirley Chisholm becomes the first black woman elected to Congress, where she is an effective proponent for minority rights and urban needs. In 1972, she campaigns for the presidency of the United States. At the Democratic national convention, she receives 151 delegate votes. She authors *Unbought and Unbossed* and receives many honors and awards. In that book, she says: "Tremendous amounts of talent are being lost to our society just because that talent wears a skirt."

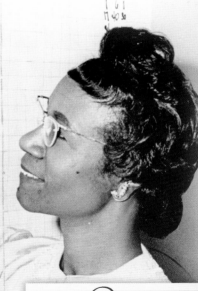

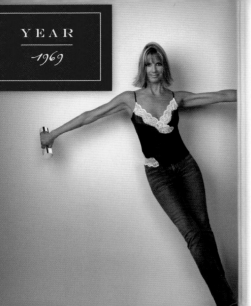

1969

Judi Sheppard Missett
helps American women stay in shape by
developing choreographed exercises set to music;
her popular exercise program is called Jazzercise.

1969

Author *Elisabeth Kübler-Ross*
writes *On Death and Dying*, defining the
emotional needs of the terminally ill. Although
she goes on to write other books, this book
becomes the standard to educate caregivers and
others on compassionate bereavement.

1969

Marilyn Horne,
opera singer extraordinaire, debuts
at La Scala in Milan, Italy.

1969

Actress and activist
Jane Fonda
is nominated for an
Oscar for *They Shoot
Horses, Don't They?*

1969

Composer and conductor *Tania León*
helps found the Dance Theatre of Harlem, and
becomes its first music director. León has received
numerous awards for her work and is recognized as
an educator and advisor to arts organizations.

1969

Coretta Scott King founds the Martin Luther King Jr. Center for Nonviolent Change.

1969

Media executive *Joan Ganz Cooney* heads a creative group to start the Children's Television Workshop, whose *Sesame Street* becomes the most successful children's television show in the history of either commercial or educational television.

1969   The National Association for the Repeal of Abortion Laws (NARAL) is founded to promote reproductive freedom and defend abortion rights.

1969

Designer *Jessica McClintock,* a widow with a young son, arrives in California with a teaching credential. She invests in a small apparel company, Gunne Sax, that she grows into an international empire known as Jessica McClintock, Inc.

1969

Golda Meir, who grows up in the United States (after emigrating from Russia), moves to Israel, and, after a long and influential political career holding a variety of posts, is elected prime minister in 1969. In her era she is the most prominent female politician in the world. She says, "Trust yourself. Create the kind of self that you will be happy to live with all of your life. Make the most of yourself by fanning the tiny, inner sparks of possibility into flames of achievement."

Her Story

African American lawyer *Eleanor Holmes Norton* is appointed chair of the New York City Commission on Human Rights, the first woman in this position. Later, she is elected to the U.S. Congress, representing Washington, D.C.

1970

1970

Well-known and beloved entertainer *Pearl Bailey* is appointed "U.S. Ambassador of Love" by President Richard Nixon. She says: "No one can figure out your worth but you."

1970

Feminist writer and activist *Kate Millett* publishes the controversial best-selling book *Sexual Politics*, which critiques the role of patriarchy in Western literature and society. In it, she says: "Many women do not recognize themselves as discriminated against; no better proof could be found of the totality of their conditioning."

1970

1970

Mountaineer *Arlene Blum* is part of the first women's climbing team to reach the summit of Alaska's Mt. McKinley (known by the locals as Denali). When team leader Grace Hoeman is taken ill, Blum, then twenty-five years old, assumes command of the six-person team. Later, Blum is the first woman to attempt to ascend Mount Everest. She says: "As long as you believe what you're doing is meaningful, you can cut through fear and exhaustion and take the next step."

1970

Photographer *Annie Leibovitz* is retained by *Rolling Stone* magazine. She quickly becomes the magazine's main photographer and wins worldwide fame for her many photographs of actors, political figures, and artists.

1970

Comanche activist *LaDonna Harris* forms Americans for Indian Opportunity; she is active in the movements for civil rights, the environment, women's issues, and world peace.

1970

Diane Crump is the first woman jockey to ride in the Kentucky Derby.

In the U.S. Congress, there are twelve women representatives and one senator. At this time, there are no female governors.

1970

JOYCE CAROL
OATES
HIGH LONESOME
NEW AND SELECTED STORIES
1966–2006

Prolific writer *Joyce Carol Oates* receives the National Book Award for her novel *Them*. Throughout her long career, she is awarded numerous honors. One of her many achievements is reinventing the genre called "Gothic fiction," which Oates uses to reimagine whole stretches of American history.

1970

1970

1970

Marie Cox founds the North American Indian Women's Association (NAIWA), the first organization to address the unique issues of its members as both women and American Indians.

1970

A group of women collaborate to form the Boston Women's Health Collective and write *Our Bodies, Ourselves*. They educate women about childbearing, sexuality, and women's health; over eight editions, the book has sold more than four million copies.

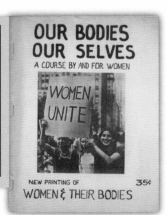

OUR BODIES
OUR SELVES
A COURSE BY AND FOR WOMEN

WOMEN
UNITE

NEW PRINTING OF
WOMEN & THEIR BODIES

35¢

1970

Ruth Westheimer earns her Ph.D. in family counseling on her way to becoming talk show host Dr. Ruth, providing frank sexual advice.

Her Story

PAGE № 165

1971

Vocalist *Carole King* releases her record album *Tapestry*. It remains on the charts for almost six years, selling more than ten million copies in the United States and an estimated twenty-two million copies worldwide. In 1990, she is inducted into the Rock and Roll Hall of Fame. She says: "Write from your heart, from your passions. If you think it's a difficult road, you're right but don't give up."

1970

Architecture critic *Ada Louise Huxtable* receives the Pulitzer Prize for distinguished criticism in the first year it is awarded.

1970

1970

Astronomer *Vera Rubin* writes about her pioneering spectroscopic research on the existence of a large percentage of dark matter in the universe.

1971

1970

Actress *Mary Tyler Moore* begins her TV show depicting a single, independent career woman.

1971

Attorney and congresswoman *Bella Abzug* helps to found the National Women's Political Caucus. She says: "We are coming down from our pedestal and up from the laundry room. We want an equal share in government and we mean to get it."

1971

1971

Physician *Helen Caldicott* becomes a spokesperson for the budding nuclear disarmament movement; she later forms Physicians for Social Responsibility, which receives the Nobel Peace Prize in 1985.

1971

Letty Cottin Pogrebin (bottom left) and *Gloria Steinem* (above) are two of the founding editors of the feminist magazine *Ms.*; it launches as a one-issue sample insert and in July 1972 becomes a monthly magazine. At that time, few realize that the magazine will become an important vehicle to declare women's rights. Women's magazines popular at the time concentrate on saving marriages, raising children, and using the right cosmetics; by contrast, *Ms.* carries articles on "desexing" the English language and discusses the then taboo subject of abortion. *Ms.* is the first national magazine to explain and advocate for the Equal Rights Amendment, to rate presidential candidates on women's issues, and to put sexual harassment and domestic violence on the cover of a woman's magazine. *Ms.* becomes immediately popular with women and generates more than twenty thousand letters from readers within the first few weeks of its publication. It is still published today. Steinem says: "If the shoe doesn't fit, must we change the foot?"

1972

Singer and entertainer *Bette Midler* releases her first record album, *The Divine Miss M*, which quickly goes gold, having sold a million copies.

1971

Jeanne M. Holm is the first woman in the U.S. Air Force to be promoted to brigadier general. Two years later, she becomes the first woman major general in all of the armed forces.

1971

1972

1972

Writer *Maya Angelou* is the first African American woman to have a feature film developed from her work, the screenplay and musical score *Georgia, Georgia*. She says: "How important it is for us to recognize and celebrate our heroes and she-roes!"

1971

Social activist *Maggie Kuhn* founds the Gray Panthers, a nonprofit organization that works for the rights and welfare of the elderly.

Patricia Schroeder is elected to Congress and arrives in Washington, D.C., with small children. She serves for twenty-four years and earns a reputation as a leader on many critical social issues. Near the end of her congressional tenure, she sees the Family and Medical Leave Act and the National Institutes of Health Revitalization Act through to becoming law. Her book about her years in Congress is titled *24 Years of House Work . . and the Place Is Still a Mess: My Life in Politics*. Her classic comment about her gender is "I have a brain and a uterus and I use both."

1972

Reporter *Linda Ellerbee* lands her first TV news position in Dallas. The president of her own production company, she has served as a network news correspondent, anchor, writer, and producer. "And so it goes" is her trademark tag line.

1972

1972

Sculptor *Eva Hesse* is the first woman to have a retrospective exhibit of her work at the Guggenheim Museum, two years after her death.

1972

Broadcast journalist *Susan Stamberg* is the first woman news anchor on a major network's national nightly news program when she begins fourteen years as the co-host of National Public Radio's highly respected newsmagazine *All Things Considered*.

Her Story

PAGE № 169

1972

Photographer, novelist, and short-story writer *Eudora Welty* wins the Pulitzer Prize for *The Optimist's Daughter*. During her long life, she is recognized as the most gifted practitioner of the short-story genre. The e-mail software Eudora is named after her in reference to her short story "Why I Live at the P.O." Welty says: "I am a writer who came of a sheltered life. A sheltered life can be a daring life as well. For all serious daring starts from within."

1972

Gertrude Boyle takes over Columbia Sportswear after being involved in the business since 1938. She grows the company from near bankruptcy to over $4 billion in annual sales.

1972

A posthumous retrospective of the work of photographer *Diane Arbus* is mounted by the Museum of Modern Art in New York City and travels through the United States, attracting more than seven million viewers.

1972

The Ford Foundation is the first major foundation to support research in women's studies with a $1 million national fellowship program for faculty and doctoral dissertation research on the roles of women in society.

1972

Juanita Kreps
is the first woman to be
appointed secretary of
commerce.

1972

"No person in the United States shall, on the basis of sex, be excluded
from participation in, or denied the benefits of, or be subject to
discrimination under any educational program or activity receiving
federal assistance." Thus reads Title IX of the Education

1972 Amendments of 1972 to the Civil Rights Act of 1964.

1972

Congress passes the Equal Rights
Amendment (ERA), which calls for all
rights under the law to be available to
both sexes. The bill needs ratification
from thirty-eight states to become
law. By 1982, time expires, and since
only thirty-five states have ratified the
amendment, it dies.

1972

Sally J. Priesand
is the first ordained female rabbi.

1973

Feminist poet and writer *Adrienne Rich* publishes *Diving into the Wreck*, which wins the 1974 National Book Award.

Film and television actress
Barbara Stanwyck
is inducted into the Hall of Great
Western Performers at the National
Cowboy and Western Heritage
Museum.

1973

1973

Physicist *Shirley Ann Jackson* is the first black woman to receive a Ph.D. from the Massachusetts Institute of Technology; later she becomes president of Rensselaer Polytechnic Institute.

1973

1973

African American lawyer
Marian Wright Edelman
sets up the Children's Defense Fund. She
says: "If you don't like the way the world
is, you change it. You have an obligation to
change it. You just do it one step at a time."

1973

Billie Jean King
is the top female athlete of
the year; she defeats male
tennis player Bobby Riggs.
She says: "No one changes
the world who isn't obsessed."

1973

9to5, a national association of workingwomen, is founded. Its mission is to strengthen women's ability to work for economic justice.

1973

Attorney *Sarah Weddington* defends a woman's right to abortion before the U.S. Supreme Court in *Roe v. Wade*.

1973

Minimalist sculptor *Anne Truitt* has a retrospective at the Whitney Museum of American Art in New York.

Her Story

1974

Radical writer in the women's movement
Andrea Dworkin publishes *Woman Hating*.

1973

Film producer *Julia Phillips* is the first woman to win an Academy Award for Best Picture as a producer for *The Sting*. Her 1977 film *Taxi Driver* is also nominated for the Best Picture Oscar.

The first shelter for battered women opens in Tucson, Arizona, offering women and their children a way to leave abusive relationships.

1973

1973

1974

1974

The U.S. Merchant Marine Academy becomes the first service academy to enroll women.

Karen Silkwood

Chemical technician *Karen Silkwood* believes her company is falsifying its records with regard to plant and worker safety. She is gathering evidence of her claim when she is killed in a mysterious car crash.

1974

1974

Katharine Meyer Graham is the first woman member of the board of the Associated Press. Later, as editor of the *Washington Post*, it is her decision to break the Watergate story. She says: "Once, power was considered a masculine attribute. In fact, power has no sex."

1974

Her Story

PAGE № 175

1974

Washington Post columnist *Mary McGrory* wins the Pulitzer Prize for commentary, for her work on the Watergate story.

1974

Entrepreneur and fierce defender of First Amendment rights *Joyce Meskis* purchases the independent bookstore Tattered Cover Books, Inc. She continues to battle the large bookstore chains for market share.

1974

1974

First Lady *Betty Ford* speaks in public about her mastectomy and urges women to get a checkup for breast cancer.

1974

Helen Thomas is the first woman to be the White House bureau chief for a news wire service; she works at the White House for over forty years.

1974

Becky Schroeder receives her first patent for an illuminated writing board (later called the Glo-sheet) while she is a preteen; she is one of the youngest Americans ever to have received a patent.

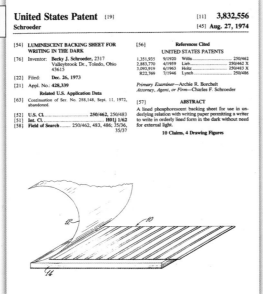

United States Patent [19]   [11] **3,832,556**
Schroeder   [45] **Aug. 27, 1974**

[54] **LUMINESCENT BACKING SHEET FOR WRITING IN THE DARK**

[76] Inventor: Becky J. Schroeder, 2317 Valleybrook Dr., Toledo, Ohio 43615

[22] Filed: **Dec. 26, 1973**

[21] Appl. No.: **428,339**

Related U.S. Application Data

[63] Continuation of Ser. No. 288,148, Sept. 11, 1972, abandoned.

[52] U.S. Cl. 250/462, 250/483
[51] Int. Cl. H01j 1/62
[58] Field of Search 250/462, 483, 486; 35/36, 35/37

[56] **References Cited**
UNITED STATES PATENTS

1,351,935	9/1920	Willis	250/462
2,883,770	4/1959	Lieb	250/462 X
3,093,919	6/1963	Holtz	250/483 X
R22,769	7/1946	Lynch	250/486

Primary Examiner—Archie R. Borchelt
Attorney, Agent, or Firm—Charles F. Schroeder

[57] **ABSTRACT**

A lined phosphorescent backing sheet for use in underlying relation with writing paper permitting a writer to write in orderly lined form in the dark without need for external light.

10 Claims, 4 Drawing Figures

1974

Well-known tennis star *Chris Evert* wins both the French Open and Wimbledon; she wins at least one Grand Slam singles title in every year from 1974 to 1986.

1975

Entertainer *Dolly Parton* receives the Country Music Association Female Vocalist of the Year Award.

1975

1975

Comedienne *Gilda Radner* launches a host of Emmy-winning characters in the television program *Saturday Night Live.* Her death from ovarian cancer at age forty-two in 1989 leads to the founding of Gilda's Club Worldwide, a support organization for people living with cancer, their family, and their friends. Radner says: "While we have the gift of life, it seems to me the only tragedy is to allow part of us to die–whether it is our spirit, our creativity, or our glorious uniqueness."

1975

Historian *Lucy Dawidowicz* publishes *The War Against the Jews 1933–1945,* a definitive study of the Holocaust.

Children's writer *Virginia Hamilton* wins the National Book Award and the Newbery Medal for *M. C. Higgins, the Great.*

1975

Golfer *Amy Alcott* wins the Orange Blossom Classic, her third tournament after turning pro, which sets the record for the fastest career win.

1975

Designer *Carole Little* founds the fashion house Carole Little, Inc. Combining casual southern California styling with chic and trendy European sportswear, the company grows from its inception in 1975 to a $500 million business.

1975

Susan Brommiller publishes a book, *Against Our Will: Men, Women and Rape,* that exposes the prevalence of violence against women.

1975

Martina Navratilova defects from Czechoslovakia and pursues her tennis career in the United States; she goes on to win a total of fifty-six Grand Slam events.

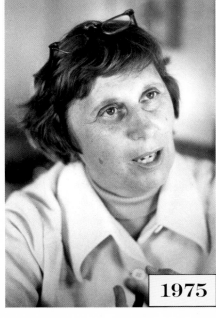

1975

Ella Grasso becomes governor of Connecticut; she is the first woman to be elected governor of a state on her own merits and not by following her husband into the office.

1975

Entertainer *Joan Rivers* wins the Georgie Award for Best Comedienne; she continues a long television career and has interviewed countless celebrities on the red carpet before the Academy Awards show.

1975

1975

Jill Ker Conway is the first female president of Smith College.

Her Story

The first issue of *Working Woman* magazine is published.

1976

1976

Maxine Hong Kingston wins the National Book Critics Circle Award for her autobiographical novel *The Woman Warrior.* She says: "To me success means effectiveness in the world, that I am able to carry my ideas and values into the world–that I am able to change it in positive ways."

1976

Deafness is not a disability for stuntwoman *Kitty O'Neil,* who races cars, jumps off buildings, and performs varied feats both on television and for the movies. She is listed in the *Guinness Book of World Records* as the "Fastest Woman on Earth."

1976

1976

Media personality *Barbara Walters* is signed as the first woman co-anchor of a major network's evening news; she continues her long career as a television commentator and celebrity interviewer. She says: "A woman can do anything. She can be traditionally feminine and that's all right; she can work, she can stay at home; she can be aggressive; she can be passive; she can be any way she wants with a man. But whenever there are the kinds of choices there are today, unless you have some solid base, life can be frightening."

To raise money for the Oneida Indian tribe's community center, *Alma Webster* and *Sandra Ninham* run a Sunday afternoon bingo game, which is the start of the Oneida reservation's multimillion-dollar gambling operation in Wisconsin.

1976

The Hyde Amendment, which prohibits federal money from being used to fund abortions for women on Medicaid, is passed.

1976

1976

Julia Robinson is the first female mathematician to be elected to the National Academy of Sciences.

1976

Conductor and opera producer *Sarah Caldwell* is the first woman to conduct at the Metropolitan Opera in New York. Caldwell says: "Learn everything you can, anytime you can, from anyone you can–there will always come a time when you will be grateful you did."

1976

Television and news anchor *Jane Pauley* is cohost of the *Today* show.

1976

Linda Alvarado owns a major Denver construction company; later she is part owner of the Colorado Rockies baseball team.

1976

Zoologist *Dixy Lee Ray* is elected Washington State's first female governor.

1976

Designer *Elizabeth Claiborne Ortenberg* founds the fashion house Liz Claiborne, Inc., which by 2006 becomes a nearly $5 billion company.

Her Story

PAGE № 184

1977

As part of the UN International Women's Year, more than two thousand delegates and eighteen thousand observers attend the government-sponsored National Women's Conference in Houston. For the opening, a lighted torch is passed 2,600 miles from Seneca Falls, New York (the birthplace of women's rights in the United States), to Houston, Texas.

1977

Kay Koplovitz launches the Madison Square Garden sports network (MSG); she later becomes the first female television network president when MSG becomes the USA Network.

1977

Josie Cruz Natori founds the high-end lingerie company Natori.

1977

Janet Guthrie is the first woman to drive in the Indianapolis 500 auto race.

1977

Physicist *Rosalyn Yalow* receives the Nobel Prize for her work in developing radioimmunoassay, a technique that uses radioactive isotopes to measure small amounts of biological substances. This technique is now in widespread use by medical professionals and scientists. She says: "We cannot expect in the immediate future that all women who seek it will achieve full equality of opportunity. But if women are to start moving towards that goal, we must believe in ourselves or no one else will believe in us; we must match our aspirations with the competence, courage and determination to succeed."

1977

Native American author *Leslie Silko* publishes her first novel, *Ceremony*, to great critical acclaim. Her writings weave oral traditions of her people with Western literary forms.

1977

Versatile actress and comedienne *Lily Tomlin* wins the first of two Tony Awards. Two of the characters she creates in the early 1970s on the television series *Rowan and Martin's Laugh-In*, Ernestine and Edith Ann, are still well known more than thirty years later.

1977

Juana Bordas (center of first row) starts the Mi Casa Resource Center for Women to advance self-sufficiency for low-income Latinas and youth.

1977

Debbi Fields founds Mrs. Fields Cookies, Inc. Fields says: "You do not have to be superhuman to do what you believe in."

1977

1977

Ann Fudge begins a long and successful executive career in marketing at General Mills; later, she becomes CEO at Young and Rubicam and the most prominent African American corporate executive on Madison Avenue.

1977

Television, theater, and film actress *Meryl Streep* appears in her first film, *Julia*. She is the most nominated actor in Academy Award history.

The Organization of Chinese American Women is founded to advocate and advance the needs of Chinese and Asian Pacific American women.

1977

Her Story

1978

Nancy Landon Kassebaum is the first woman to serve in the U.S. Senate having been neither elected to serve first in the House of Representatives nor appointed to fill out the remainder of a term from a husband after his death while in office.

1978

Faye Wattleton is elected the president of Planned Parenthood; she is the first African American and youngest person to be elected president.

Maude de Victor is a navy veteran and a veterans' benefits counselor who makes the link between the spraying of Agent Orange in Vietnam and veterans' health problems and cancer-related illnesses.

1978

1978

Electrical engineer *Judith Resnik* is selected for NASA's astronaut program and serves as a mission specialist aboard the space shuttle *Discovery*. In 1986, at age thirty-six, she perishes along with the entire crew of the space shuttle *Challenger*.

The U.S. Army integrates women into the formerly all-male body, dismantling the Women's Army Corps, which began during World War II.

1978

Women Against Violence in the Media do a "Take Back the Night" march to draw attention to a woman's right to walk at night without fear and to honor and acknowledge women survivors of domestic violence, sexual assault, and other forms of violence against women.

1978

1978

Lakota Indian educator *Patricia Locke* advocates for the American Indian Religious Freedom Act, which sets into federal law the right of Native Americans to freely practice their spiritual traditions. She helps organize seventeen tribally run colleges and is awarded a MacArthur Foundation Fellowship in 1991.

1979

Feminist artist *Judy Chicago* completes a huge art project called "The Dinner Party" and begins to tour the country with it.

Her Story

PAGE Nº 185

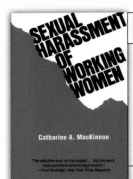

1979

Catharine Alice MacKinnon, a groundbreaking feminist attorney, publishes *Sexual Harassment of Working Women,* which argues that sexual harassment is a form of sexual discrimination under Title VII of the Civil Rights Act of 1964.

1979

Designer *Adrienne Vittadini* begins her clothing firm as a hobby; it grows into a multimillion-dollar industry targeting professional women.

1979

Patty DeDominic founds PDQ Personnel Services, one of California's largest independent staffing firms, with offices nationwide. By 2004 she is the CEO of PDQ Careers Group and has long been recognized by Congress, business, and labor groups as a human resources expert.

To improve conditions for women, Lakota women form the Sacred Shawl Society, an intervention program to deal with domestic violence within a cultural context.

1979

1979

Patricia Roberts Harris is the first woman to be named to two cabinet positions: secretary of housing and urban development in 1979 and secretary of health and human welfare in 1980. At her confirmation hearing she was asked if she could understand the needs of the poor. She replies: "Senator, I am one of them. You do not seem to understand who I am. I am a black woman, the daughter of a dining-car worker. . . . If my life has any meaning at all, it is that those who start out as outcasts can wind up as being part of the system."

1979

Science fiction writer *Octavia Butler* publishes *Kindred,* one of her best-selling novels. She is an inspiration and model for other black writers who write in the science fiction and fantasy genre.

The problems associated with domestic violence gain national attention when the National Coalition Against Domestic Violence holds its first conference.

1980

Called the "Prison Angel," *Sister Elaine Roulet* is a crusader for the children of women in prison; she founds and is executive director of Providence House, Inc.

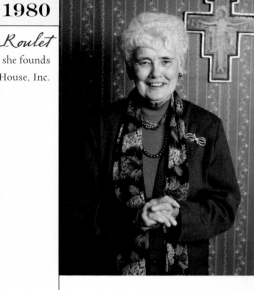

1980

Sherry Lansing is appointed president of 20th Century Fox. She becomes the first woman president of a major Hollywood studio.

The National Women's History Research Project (later called the National Women's History Project) is established. Participants lobby successfully first for National Women's History Week and then for National Women's History Month (March).

1980

A Lot of Teenagers Are Dying For A Drink

DRINKING DRIVERS ARE INVOLVED IN ONE HALF OF ALL FATAL ACCIDENTS IN CALIFORNIA

DRINKING DRIVERS ARE INVOLVED IN ONE FOURTH OF ALL INJURY ACCIDENTS IN CALIFORNIA

MORE THAN 275,000 DRINKING DRIVERS ARE ARRESTED AND TAKEN TO JAIL EACH YEAR IN CALIFORNIA

To all this we say... Enough!!

...ER, DRIVING UNDER THE INFLUENCE LAWS ALSO ...UDE UNDER THE INFLUENCE OF ANY PRESCRIBED OR ...LLEGAL DRUG ALONE, OR IN COMBINATION WITH ALCOHOL.

1980

1980

After her daughter is killed by a drunk driver, *Candy Lightner* organizes Mothers Against Drunk Driving (MADD).

Her Story

1981

1981

Attorney and judge *Sandra Day O'Connor* becomes the first woman to sit on the U.S. Supreme Court. In 1952, when she finishes her law degree at Stanford, graduating third in her class, no California law firm is willing to hire her because she is a woman. O'Connor says: "As the first woman nominated as a Supreme Court Justice, I am particularly honored. But I happily share the honor with millions of American women of yesterday and today whose abilities and conduct have given me this opportunity of service."

1981

In 1981, *Gerda Lerner* is the first woman in fifty years to become president of the Organization of American Historians; she is acknowledged as one of the foremost scholars in the field of women's history. Dr. Lerner has published ten books on women in history, encompassing topics such as the Grimké sisters and the need to eliminate the invisibility of women. She says, "We stand at the beginning of a new epoch in the history of humankind's thought, as we recognize that . . . woman, like man, makes and defines history."

1981

Wilhelmina Holladay opens the National Museum of Women in the Arts in Washington, D.C.

1981

At age twenty-one, sculptor *Maya Lin* is selected to design the Vietnam Veterans Memorial; she wins international acclaim for her site-specific art and architectural projects.

1981

Producer and Emmy winner *Marcy Carsey* founds the production studio Carsey Werner; she is later named one of the fifty greatest women in radio and television.

1981

Writer and human rights advocate *Bette Bao Lord* writes her first novel, *Spring Moon*, which becomes an international best seller and American Book Award nominee for best first novel. A Chinese immigrant, Bao Lord has continued to write both novels and nonfiction, and her writings have been translated into fifteen languages. She also lectures frequently on foreign affairs.

1981

Political scientist *Jeane Kirkpatrick* is the first female U.S. delegate to the United Nations. She says: "Democracy not only requires equality but also an unshakable conviction in the value of each person, who is then equal."

1981

Inventor *Anne Chiang* is best known for her work on low-temperature glass-compatible polysilicon coatings for large-area electronics, otherwise known as thin-film transistor technology, or TFT. Today's flat-panel-display personal computers are a direct result of her work, which also enables cell phone and other portable communications systems.

Her Story

PAGE №. 189

1982

Rebecca Matthias (right) founds MothersWork, Inc., a firm that makes fashionable maternity clothes.

Writer *Alice Walker* publishes the Pulitzer Prize–winning novel *The Color Purple;* she is the first black woman to win this award.

1982

1982

1983

1982

After her first album is released in 1964, Cree Indian singer-songwriter *Buffy SainteMarie* receives many honors and awards. She wins an Academy Award in 1982 for her song "Up Where We Belong," which is the theme song for the movie *An Officer and a Gentleman.* Sainte-Marie's talents are in many fields; she appears on *Sesame Street* for five years with her son, she operates the Nihewan Foundation for Native American education, and she earns a Ph.D. in fine art.

Standup comic and talk show host *Ellen DeGeneres* is selected as the funniest woman in America by the cable channel Showtime.

1982

1982

Maria de Lourdes Sobrino establishes Lulu's Desserts to manufacture and distribute ready-to-eat gelatin-based desserts; it is now a multimillion-dollar business.

Chicana academics found Mujeres Activas en Letras y Cambio Social (MALCS): Women Active in Leaders and Social Change to support Chicana/Latina and Native American students and faculty and to fight race and gender discrimination in higher education.

1982

1982

Nancy Brinker (on left) founds the Susan G. Komen Breast Cancer Foundation in memory of her sister, who died of breast cancer in 1984 at age thirty-six. One of Brinker's early funding strategies is the Race for the Cure, at that time a unique way to raise money for charity. The Komen Foundation continues to raise money and awareness of breast cancer and supports research aimed at its eradication.

1983

Violinist *Ellen Zwilich* is the first female composer to win the Pulitzer Prize in music; later she is named the first-ever occupant of Carnegie Hall's Composer's Chair. Her music is described as having "fingerprints," immediately recognizable as the product of a particular composer.

1983

Jenny Craig founds the weight loss and nutritional food company Jenny Craig, Inc.

Her Story

1983

Dancer, choreographer, anthropologist, and writer *Katherine Dunham* is a Kennedy Center honoree. She takes dance in new directions by combining modern and ballet techniques.

1983

Connie Chung is the first Asian American to anchor a national newscast at a major network. She says: "I don't know what all of us can do to continue to press for more women, more minorities, but it's just something that we all have to work on."

1983

Esther Dyson is the chair of EDventure Holdings, a company focusing on information technology worldwide. Dyson is well respected for her insights into industry trends. She writes frequent articles on emerging digital technology and insightful analyses of such issues as the impact of the Internet on the U.S. education system.

1983

Broadcast journalist *Lesley Stahl* begins hosting the television news program *Face the Nation.*

Elizabeth Hanford Dole is appointed Secretary of Transportation; she later serves in a second administration as Secretary of Labor. After heading the Red Cross, she is elected to the Senate. On her career aspirations, Dole says: "Women share with men the need for personal success, even the taste for power, and no longer are we willing to satisfy those needs through the achievements of surrogates, whether husbands, children or merely role models."

1983

Sally Ride is in the first group of American women astronauts; she is the first American woman to orbit the earth while aboard the space shuttle *Challenger*.

1983

1983

In 1983, photographer **Marion Ettlinger** refocuses her career to concentrate on taking pictures of book authors, a very specialized niche. Ettlinger's work appears in more than six hundred books. In 2003, she collects two hundred portraits of contemporary writers for a coffee table book, *Author Photo*.

1983

Geneticist **Barbara McClintock** is the first woman to win an unshared Nobel Prize in physiology or medicine for her discovery of the ability of genes to change positions on the chromosome (jumping genes).

Her Story

1984

Catherine Crier is the youngest elected state judge in Texas history; later she is an award-winning journalist and Court TV host.

1984

Molecular biologist *Flossie Wong-Staal* is the first person to clone the HIV virus, a major research advance in the treatment of AIDS.

1984

Designer *Donna Karan* founds the fashion house Donna Karan.

1984

1984

Novelist, poet, and MacArthur Fellowship recipient *Sandra Cisneros* publishes her first work of fiction, *The House on Mango Street*. Much of her work is inspired by her Mexican American heritage.

1984

Runner *Joan Benoit* wins the gold medal in the first-ever women's marathon at the Summer Olympics; she wins by more than a minute over her nearest rivals.

1984

Broadcast journalist *Diane Sawyer* becomes the first woman reporter on television's *60 Minutes*.

1984

1984

Geraldine Ferraro is named Democrat Walter Mondale's running mate. This is the first time a major American political party has nominated a woman for the vice presidency.

Her Story

1985

Harvard Business School professor and author *Rosabeth Moss Kanter* is one of the most prominent business speakers and strategy consultants in the world. Her focus for more than twenty-five years is helping to guide organizations and their leadership through change.

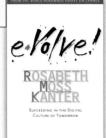

1985

Libby Riddles is the first woman to win the famous Iditarod Trail Sled Dog Race in Alaska.

1985

Wilma Mankiller becomes the first woman principal chief of the Cherokee Nation. She remains in office until 1995, during which time she works to bring back balance and reinvigorate the Cherokee Nation through community-building projects. She helps to bring about improved health care, education, utilities management, and tribal government. Mankiller says: "A lot of young girls have looked to their career paths and have said they'd like to be chief. There's been a change in the limits people see."

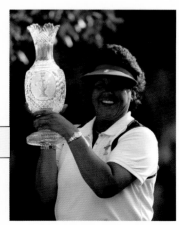

1985

Golfer *Nancy Lopez* is named player of the year and AP Athlete of the year and wins the Ladies Professional Golf Association championship.

1985

Vinita Gupta founds the company Digital Link (now Quick Eagle Networks); the company becomes significant in the data telecommunications industry.

Multitalented *Oprah Winfrey* begins her talk show, *The Oprah Winfrey Show*. Actress, producer, director, talk show host, editorial director of *O: The Oprah Magazine*, and cofounder of women's cable network Oxygen Media, she exerts much influence on the American TV viewing public. About her philanthropy and generosity with her time, energy, and money, she says, "My whole quest at this point in my life . . . is: How do I use my life, my whole life, that includes: my money, my resources, my access, my so-called perceived influence–How do I use my whole self in order to bring goodness and some light into the world?" In acknowledging the women who came before her, she says: "I have crossed over on the backs of Sojourner Truth, Harriet Tubman, Fannie Lou Hamer, and Madam C. J. Walker. Because of them I can live the dream. I am the seed of the free, and I know it. I intend to bear great fruit."

1985

Astronaut *Shannon Lucid* begins the first of her five space flights. She holds the U.S. single-mission space flight endurance record (188 days) on the Russian space station *Mir*.

The New York Asian Women's Center sponsors programs to combat violence against Asian women.

1985

1985

1986

Singer and songwriter *Gloria Estefan* popularizes Latin music when "Conga" becomes an American pop hit single. The five-time Grammy Award winner has sold more than ninety million record albums worldwide.

1986

Nien Cheng publishes her autobiographical book *Life and Death in Shanghai*.

1986

1986

Pleasant T. Rowland begins a company making historically accurate dolls that appeal to seven-to-twelve-year-olds. She adds books, accessories, and *American Girl* magazine, all of which are sold in 1998 to toy company Mattel, Inc., for $700 million.

1986

Susan Butcher wins the first of four Iditarod Trail Sled Dog races in Alaska. Butcher says: "I do not know the word 'quit.' Either I never did it, or I have abolished it."

1986

Outspoken television talk show host, actress, and comic *Rosie O'Donnell* begins her television career on *Gimme a Break!*

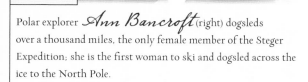

Polar explorer *Ann Bancroft* (right) dogsleds over a thousand miles, the only female member of the Steger Expedition; she is the first woman to ski and dogsled across the ice to the North Pole.

1986

Christa McAuliffe, a social studies teacher, is chosen to be America's first private citizen in space; during the mission all crew members perish aboard the space shuttle *Challenger*.

1986

Television producer *Caryn Mandabach* is part of a trio of talented producers that includes Tom Werner and Marcy Carsey. Mandabach works with Norman Lear and goes on to take comedian Bill Cosby's vision for a family show and create the most successful sitcom up until then. In 1986, she wins an Emmy for her work on *The Cosby Show*. Mandabach is also part of the creative team that develops another successful television comedy series, *3rd Rock from the Sun*, which runs from 1996 through 2004.

1986

Four-time women's world chess champion *Susan Polgar* is a pioneer in women's chess and a true role model to millions of women and children. She is the first woman to earn the men's grandmaster title and the first woman to qualify for the men's world championship.

Her Story

1987

Gloria Anzaldúa writes *Borderlands/La Frontera*; it is chosen as one of the best new books of the year by the *Literary Journal*.

1987

1987

Educator, humanitarian, and women's rights activist *Johnetta B. Cole* is the first female president of Spelman College. Cole says: "When you educate a man you educate an individual, but when you educate a woman, you educate a nation."

1987

Katherine Siva Saubel, a preserver of tribal culture and languages, is elected Elder of the Year by the California State Indian Museum.

President Ronald Reagan selects *Jo Waldron,* who is deaf, to receive the President's Trophy; she is appointed "Disabled American for the Nation" to represent all Americans with disabilities. She works for eleven years to ensure passage of the Americans with Disabilities Act of 1990.

1987

1987

Camille Olivia Hanks Cosby and her husband, Bill Cosby, provide tremendous support to black colleges; in 1987 they give $20 million to Spelman College.

Women in Franchising forms to provide training in this area.	**1987**

1987

Computer scientist *Anita Borg* founds the Systers e-mail list to provide mentors, support, encouragement, and information sharing to women in computing.

Her Story

PAGE № 201

Democratic politician *Nancy Pelosi* is elected to the U.S. House of Representatives from California; later she is the first woman to lead a major political party in either house of Congress. With the election of 2006, she becomes the first female Speaker of the House, about which Pelosi says: "It's an historic moment for the Congress, it's an historic moment for the women of America. It is a moment for which we have waited over 200 years. Never losing faith, we waited through the many years of struggle to achieve our rights. But women weren't just waiting, women were working, never losing faith we worked to redeem the promise of America, that all men and women are created equal. For our daughters and our granddaughters, today we have broken the marble ceiling. For our daughters and our granddaughters, the sky is the limit. Anything is possible for them."

1987

Physician *Mae Jemison* is the first black woman to be selected as an astronaut. She says: "Don't let anyone rob you of your imagination, your creativity, or your curiosity. It's your place in the world; it's your life. Go on and do all you can with it, and make it the life you want to live."

1987

Sheri Poe starts the Ryka Company, which sells athletic shoes for women and gives a portion of the proceeds to cancer research.

1987

Scholar, teacher, and administrator *Donna Shalala* is named chancellor of the University of Wisconsin–Madison; she is the first woman to preside over a Big Ten school. In 1993, she is named secretary of health and human services.

1987

Singer *Aretha Franklin* is described as the "Queen of Soul"; she is the first woman inducted into the Rock and Roll Hall of Fame.

1988

Ann Richards, the former governor of Texas, delivers the keynote address to the Democratic national convention.

1988

Television writer and producer *Diane English* has several successful series to her credit; *Murphy Brown* is among them. During the five years of *Murphy Brown*, English wins three Emmy awards and receives five additional Emmy nominations. In *Murphy Brown* and her other shows produced before and after it, English explores how career women adapt themselves to work in a professional world that is dominated by men.

1988

1988

Marin Alsop wins the Leonard Bernstein Conducting Fellowship; she continues to win awards from all over the world for her fresh, innovative conducting. The first conductor to be named a MacArthur Fellow, Alsop is also the first woman to head a major American orchestra when she is named the music director of the Baltimore Symphony in 2007–8.

1988

Athlete *Florence Griffith Joyner* sets a world record in the 200-meter sprint at the Olympics and wins three gold medals and one silver medal.

1988

Penny Marshall directs *Big*, the first film directed by a woman to gross more than $400 million at the box office.

Biochemist *Gertrude Elion* wins the Nobel Prize for a leukemia-fighting drug. She holds forty-five patents; later she is inducted into the National Inventors Hall of Fame and receives the National Medal of Science.

Bonnie Blair is the first American to win gold medals in three straight Olympic games and the first American woman to win five gold medals in the history of the Olympics.

Her Story

PAGE № 205

1989

Sarah Brady becomes a leader in the gun control movement after her husband, Jim Brady, is permanently injured during an attempted assassination of President Ronald Reagan.

1988

Athlete *Jackie Joyner-Kersee* wins Olympic gold medals in the heptathlon in 1988 and 1992.

1989

Prolific writer *Amy Tan* publishes *The Joy Luck Club*. The daughter of Chinese immigrants, Tan serves as co-producer and co-screenwriter for the film adaptation of her novel.

1988

1989

1988

Debi Thomas is the first black woman to win a figure skating medal at the Winter Olympics.

1989

Amy Domini creates the Domini 400 Social Index; later she starts the Domini Social Equity Fund, the oldest socially and environmentally screened index fund in the United States.

Playwright *Wendy Wasserstein* enjoys big success with *The Heidi Chronicles*, which describes women's changing roles.

1989

Businesswoman *Steffie Allen* founds the AthenA Group; she later begins the nonprofit Women'sVision Foundation to transform the American workplace to become more supportive of all employees.

1989

Julia Chang Bloch is the first Asian American to become a U.S. ambassador, to Nepal; she later is the president of the nonprofit organization U.S.-China Education Trust.

1989

1989

Choreographer, dancer, and artistic director *Judith Jamison* is the first black woman to head a major dance company.

Her Story

PAGE № 207

1990

Breast cancer surgeon *Susan Love* discusses issues related to women's health; *Dr. Susan Love's Breast Book* becomes the standard reference text in the field. She establishes the Dr. Susan Love Research Foundation, whose mission is to eradicate breast cancer and improve the quality of women's health through innovative research, education, and advocacy.

1990

Author *Sally Helgesen* writes *The Female Advantage*, which becomes a best seller; it identifies what skills women have to contribute to businesses, rather than how women need to adapt and change to suit others.

1990

Author and lecturer *Anne Wilson Schaef* writes and speaks about women needing to find more balance in their lives and to take time for themselves.

1990

Marine biologist and oceanographer *Sylvia Alice Earle* is the first woman to serve as chief scientist at the National Oceanic and Atmospheric Administration (NOAA).

1990

Comedienne and actress *Whoopi Goldberg* wins an Academy Award for her work in *Ghost*; she is the first black woman to win since Hattie McDaniel in 1939.

1990 Women at Brown University begin a graffiti campaign to publicize the names of male students who commit date rape.

1990 Physician *Antonia Novello* is the first woman and first Latina appointed U.S. surgeon general.

1990

1990 Broadcast journalist and television personality *Katie Couric* joins the *Today* show as cohost. In 2006 she becomes anchor of the *CBS Evening News*.

1990 Electrical engineer *Ellen Ochoa* is the first Hispanic woman astronaut; she also invents optical analysis systems.

1991

Writer and activist *Audre Lorde* (Gamba Adisa) is named poet laureate of New York State.

1991

News analyst and congressional correspondent *Cokie Roberts* wins an Emmy Award for her ABC News special *Who Is Ross Perot?*; later she writes *Founding Mothers: The Women Who Raised Our Nation.*

1991

Cardiologist *Bernadine Healy* is the first woman to head the National Institutes of Health.

~1991~

1991

Pulitzer Prize–winning journalist and author *Susan Faludi* publishes *Backlash*, documenting conservative reactions against women's rights and advancement.

1991

Marilyn VanDerbur Atler, a former Miss America, goes public to tell her story of incest.

Washington, D.C., elects its first woman mayor, *Sharon Pratt Dixon Kelly;* she is the first black woman to lead one of the top twenty cities in the country.

1991

The allegations of sexual harassment made by *Anita Hill* against Supreme Court nominee Clarence Thomas raise national awareness about the issue of sexual harassment. Hill says: "I am really proud to be a part in whatever way of women becoming active in the political scene. I think it was the first time that people came to terms with the reality of what it meant to have a Senate made up of 98 men and two women."

1991

Classical composer and pianist *Shulamit Ran* wins the Pulitzer Prize in music.

1991

Jodie Foster, already a well-known actress, demonstrates her talent in the male-dominated field of directing with *Little Man Tate,* her directorial debut.

1991

Debora de Hoyos is the first woman named managing partner at one of the nation's largest U.S. law firms.

1991

Her Story

1992

Author *Gail Sheehy* publishes her groundbreaking book *The Silent Passage: Menopause*, one of her books on the stages of life.

Carol Bartz is chair and CEO of Autodesk, Inc., the world's leading supplier of design software.

1992

1993

Author and women's advocate *Sheila Wellington* is president of Catalyst. Focusing first on the census of women on corporate boards, Wellington grows the organization to become the preeminent U.S. nonprofit focused on women's issues.

1992 Fifty-eight percent of women with children under the age of six are in the paid workforce.

1992

1993

1992

The most moving speech of the Republican national convention is given by *Mary Fisher,* an HIV-positive mother; since that time she continues to be an international speaker and activist focusing on the AIDS epidemic.

1993

Jocelyn Elders is selected as the surgeon general of the United States.

1993 *Ms.* magazine sponsors the first Take Our Daughters to Work Day on the fourth Thursday in April.

1993

Harvard Law School graduate *Janet Reno* is selected as the first female U.S. attorney general.

1993

Best known for romantic comedies, novelist *Nora Ephron* writes the screenplay and directs the film *Sleepless in Seattle*.

1993

Julie Krone is the first woman jockey to win a Triple Crown horse race; later she becomes the first female jockey in the National Thoroughbred Hall of Fame.

1993

Judge *Ruth Bader Ginsburg* is the second woman to serve as an associate justice on the U.S. Supreme Court. When she graduates first in her class from Harvard Law School in 1959, no law firm will hire her, as she is Jewish and a mother. Ginsburg says: "I pray that I may be all that {my mother} would have been had she lived in an age when women could aspire and achieve and daughters are cherished as much as sons."

1993

Pulitzer Prize-winning *Rita Dove* is the first black woman and youngest person to be named poet laureate of the United States.

1993

Her Story

1993

Aeronautical engineer
Sheila Widnall
is the first woman to head
one of the country's military
branches when she becomes
secretary of the air force.

1993

The Vietnam Women's
Memorial, designed by
Glenna Goodacre,
is unveiled.

1993 The Texas Rangers, a mounted fighting force and an elite corps of state troopers founded in 1835, admit two women, Marrie Reynolds Garcia and Cheryl Campbell Steadman.

1993

1993

Political scientist
Condoleezza Rice
becomes the first black chief
academic officer at Stanford when
she is named provost; later Dr. Rice
becomes national security advisor in
one presidential administration and
secretary of state in another.

1993

Toni Morrison wins the Nobel Prize in literature for her body of work; she is the eighth woman as well as the first black to win. She says: "If there's a book you really want to read, but it hasn't been written yet, then you must write it."

1994

Myra Sadker publishes *Failing at Fairness: How America's Schools Cheat Girls*; later the nonprofit Myra Sadker Advocates is established, dedicated to promoting gender equity in and beyond schools.

1994

Skier *Picabo Street* wins an Olympic downhill silver medal.

1994

1994

Humorous entertainer *Fran Lebowitz* releases *The Fran Lebowitz Reader*.

Her Story

PAGE № 245

1995

Eileen Collins becomes the first American woman shuttle commander when she pilots the space shuttle *Discovery*.

1995

Ann Livermore is elected as a corporate vice president of Hewlett-Packard. Later, after helping to devise HP's first Internet strategy, she is promoted to CEO of the company's $14 billion Enterprise Computing Division.

1995   1996

1995 The Glass Ceiling Commission reports that 95 to 97 percent of the senior management positions in corporate America are held by white men.

1995

Ruth Simmons is the first black woman to lead a top university, Smith College; in 2004 she leaves Smith to be the first female and first black president at Brown University.

1995

Linda Chavez-Thompson is elected the executive vice president of the AFL-CIO. She says: "The face of labor is changing, and you can tell this by the mere fact that I am a woman . . . and a woman of color."

1995

Serena Williams becomes a professional tennis player at age fourteen. She says: "Luck has nothing to do with it, because I have spent many, many hours, countless hours, on the court working for my one moment in time, not knowing when it would come."

Audiologist *Marion Downs* establishes the Marion Downs National Center for Infant Hearing at the University of Colorado at Boulder. Universal Newborn Hearing Screen (UNHS), which she has advocated for fifty years, is now becoming the standard of care throughout the world.

1996

Athlete, broadcaster, and sports commentator *Robin Roberts* (left) is honored with the creation of a sports journalism scholarship. Later, she joins the television team at ABC's *Good Morning America* as its third anchor.

1995

Because of her athletic prowess, basketball player *Sheryl Swoopes* is the first female athlete to have a shoe named after her (Nike Air Swoopes).

Her Story

PAGE № 217

Half-Life of a Zealot

Swanee Hunt

1997

Philanthropist and former ambassador *Swanee Hunt* is the founding director of the Women and Public Policy Program at Harvard's Kennedy School of Government.

 1997

Original artist, sculptor, and director *Julie Taymor* is the first woman to receive a Tony Award for directing a musical, *The Lion King.* She later wins two Academy Awards for the movie *Frida*, which she directs.

1997

Advertising executive *Shelly Lazarus* is the CEO of Ogilvy and Mather Worldwide.

1997

Entrepreneur *Martha Stewart* buys back her company from Time Warner and relaunches it as Martha Stewart Living Omnimedia, Inc.

1997

Violet Palmer, a former recreation administrator, is selected by the National Basketball Association as the first black woman referee in a male sports league.

1997

Jody Williams wins the 1997 Nobel Peace Prize for her work to eliminate antipersonnel land mines; she founds the International Campaign to Ban Landmines (ICBL).

1997 Diplomat *Madeleine Albright* is sworn in as the seventy-fourth secretary of state, the first female and highest-ranking woman in the history of the U.S. government, a position she holds until 2001. Since that time, Dr. Albright has published her memoirs (*Madam Secretary: A Memoir*) and an additional book entitled *The Mighty and the Almighty*. She serves as chair of the National Democratic Institute for International Affairs.

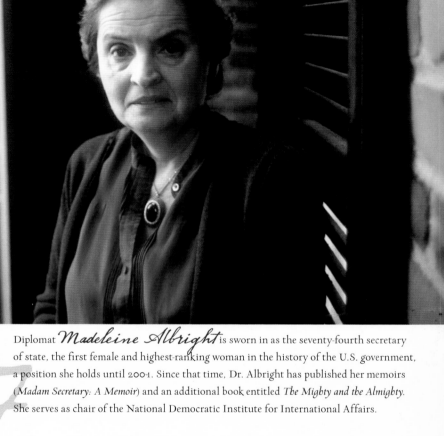

1997 Lois Hart, Marilyn Laverty, Linda Rydberg, Charlotte Waisman, Lisa Niederman, and others begin meeting regularly to create the Women's Leadership Institute.

1997

Jill Barad becomes the CEO of toy company Mattel, Inc., after working her way up through the company; she has headed the Barbie doll line since 1983.

Her Story

1998

Highly regarded Wall Street analyst *Abby Joseph Cohen* makes partner at the investment firm of Goldman Sachs. She is their chief U.S. portfolio strategist and is dubbed the "Prophet of Wall Street" for her insightful predictions of stock market growth.

1999

Choreographer and founder of a predominantly black, internationally recognized dance ensemble, *Cleo Parker Robinson* is named to the National Council of the Arts.

1998

Meg Whitman is the president and CEO of eBay, an online marketplace for goods and services.

1998

Wilma Webb is appointed the U.S. Labor Department secretary's representative for the six states encompassing the Rocky Mountains, the first woman to serve in this position.

 1998

1999

1999

Mia Hamm, who began her international athletic career at age fifteen, scores her 108th international goal to break a record formerly held by an Italian; in the same year she helps the U.S. team win soccer's World Cup.

1999

Carly Fiorina is named president and CEO of Hewlett-Packard, the first woman to hold that position. She will remain in that post until 2005.

1998

Known for her success at Nickelodeon, a television cable network targeted to children, *Geraldine Laybourne* and others found the cable television network, Oxygen Media. It caters to the interests of women.

Cathy L. Hughes is the founder and owner of Radio One, the first radio chain to target the black market. In 1999, when she takes the station public, it is the largest black-owned and -operated broadcast company in the nation. She is the first black woman with a company on the American Stock Exchange; at that time its value is in excess of $2 billion.

1999

Andrea Jung, who began with Avon Products, Inc., in 1994, is elected the company's CEO. She is responsible for Avon's long-term strategic development.

2000

Susan Decker, with a background in equity research, joins Yahoo!; later she becomes CFO and executive vice president. She plays a key role in determining Yahoo!'s business strategy.

2000

2000

Former First Lady *Hillary Rodham Clinton* is elected to the U.S. Senate.

2000   The Women's Museum: An Institute for the Future opens in Dallas, Texas. The nation's only comprehensive women's history museum is a Smithsonian Institution affiliate.

Her Story

PAGE № 224

2001

Lawyer and former Colorado Attorney General *Gale Norton,* a lifelong conservationist, is named U.S. secretary of the interior.

2001

After working for the company in various capacities for more than twenty years, *Colleen C. Barrett* is named president and corporate secretary for Southwest Airlines.

2001

Regina Carter, who begins playing the violin at age four, is the first jazz musician to play the 250-year-old Guarneri violin once owned by Paganini.

200·1

2002

Engineer *Helen Greiner* invents the Roomba robot vacuum and co-founds iRobot Corporation. Under her leadership, the company grows from a garage venture to over $50 million.

2001

Former New Jersey governor (the first woman to hold that position) *Christine Todd Whitman* is appointed administrator of the Environmental Protection Agency.

2002

Journalist *Karen Elliot House* is named publisher of all print editions of the *Wall Street Journal*. She is a senior vice president of Dow Jones and Company and a member of the executive committee.

2002

Anne M. Mulcahy, who begins with the Xerox Corporation in 1976, is named its chairman and CEO.

2002

Patricia Russo is named chair and CEO of Lucent Technologies.

2002

2002

Ann Moore, who joined Time, Inc., in 1978 as a financial analyst, is appointed chair and CEO of the company, which is the world's largest publisher of magazines.

2002

Halle Berry is the first woman of African American descent to win the Oscar for Best Actress, for *Monster's Ball*.

Her Story

PAGE № 223

2003

Sharon Allen is selected as board chair of Deloitte and Touche USA; she is the highest-ranking woman in the firm's history and the first woman to be chair at a leading professional services firm. She is responsible for the governance of an $8 billion corporation.

2003

Robotics research engineer *Ayanna Howard* is named one of the world's top young innovators by *Technology Review*; she works at the Telerobotics Research and Applications Group at the Jet Propulsion Laboratory in California.

2003

After thirteen years with Wal-Mart Stores, Inc., *Linda Dillman* is promoted to executive vice president and chief information officer. She is instrumental in developing a sophisticated information service network, a key component of Wal-Mart's competitive business strategy.

2003

Jane Friedman, who in 1997 was named president and CEO of HarperCollins Publishers after a long career at Random House, is selected as chair of the Association of American Publishers; she is only the second woman to hold this post.

2003

Mary Sammons is selected as president and chief executive officer of Rite Aid Corporation. One of a small but growing number of women CEOs, Sammons has twenty-six years of experience at a general merchandise retailer, where she worked her way up in positions of increasing responsibility.

The Nobel Prize in physiology or medicine is awarded to *Linda B. Buck* for her pioneering work, which clarifies an understanding of the human sense of smell.

2004

Financial analyst *Sallie Krawcheck* is appointed CFO and head of strategy at Citigroup, Inc.

2004

Janet L. Robinson, who came to the New York Times Company in 1983, is selected as its president and CEO.

2004

With more than twenty years of U.S. and international experience, *Susan Ivey* is named chairman and CEO of Reynolds American Inc. Reynolds is the nation's second largest cigarette company. Ivey serves on the Women's Leadership Initiative for the United Way of America and on the Committee of 200, an international organization of women CEOs, entrepreneurs, and business leaders who provide mentoring, education, and support for aspiring women in business.

2004

2004

Carol Kovac, who has worked at IBM for more than twenty years and held many management positions, launches an emerging business unit, becoming general manager of Healthcare and Life Sciences at IBM. It quickly becomes a multibillion-dollar business and one of the most successful IBM ventures to date, with more than fifteen hundred employees worldwide.

2004

Ann Sarnoff is selected as the chief operating officer (COO) of the Women's National Basketball Association after serving as COO of VH1 and CMT: Country Music Television. In 2006, she leaves to become president of Dow Jones Ventures.

Her Story

2004

After more than twenty years of
helping to build a music-video channel,
Judy McGrath becomes
chairman and CEO of MTV Networks.

2004

2004

Experienced television, film, and publishing executive
Susan Lyne is named chief executive of Martha
Stewart Living Omnimedia.

2004

Distinguished neuroscientist *Susan Hockfield*
is named president of the Massachusetts Institute of Technology
(MIT), the first woman to hold the job.

2004

Organic chemist *Stephanie Burns*
is named the first woman president and CEO of
Dow Corning.

2004

Phylicia Rashad is the first black actress to win a
Tony Award. She receives the award for her performance in the
revival of Lorraine Hansberry's play *A Raisin in the Sun*.

2005

After working in the White House kitchen for ten years as an assistant, *Cristeta Comerford* is selected as head chef, the first woman in this top position.

2004

Biochemist *Julie Theriot* wins a MacArthur Fellowship. Her research in the field of bacterial infection aims to make our food supply safer. She focuses on two foodborne bacteria: shigella, which causes dysentery, and listeria, which can be very dangerous for pregnant women, newborns, and people with weakened immune systems.

2005

2005

Carolyn Vesper Bivens is named commissioner of the Ladies Professional Golf Association and is the first woman to lead the organization in its fifty-five-year history.

2005

Brenda Barnes is named president and CEO of Sara Lee Corporation. In 1998, she surprised the business world and blazed a trail for women when she stepped down as CEO of PepsiCo to spend time with her family.

Her Story

PAGE № 227

Paula Rosput Reynolds is named Safeco's new president and chief executive officer. Reynolds' prior experience includes twenty-seven years in the energy business and five years' tenure as CEO of AGL Resources, an Atlanta-based holding company.

2006

The Wall Street industry watchdog organization, the National Association of Securities Dealers (NASD), selects *Mary Schapiro* as its chair and chief executive, the first woman to head the sixty-six-year-old organization. Schapiro is a former commissioner at the Securities and Exchange Commission and was chair of the Commodities Futures Trading Commission. NASD in 2005 brought a number of disciplinary actions and collected a record $125.4 million in fines.

2006

2006

The president and CFO of PepsiCo, *Indra K. Nooyi,* is named CEO. This promotion ranks her as number two among the eleven women CEOs in the Fortune 500.

2006

Patricia Woertz is appointed the president, chief executive officer, and member of the board of directors of Archer Daniels Midland Company. While in her previous position as executive vice president of Chevron Corporation, Woertz focused on one of her personal passions—the well-being of employees. During her tenure she reduced injuries by more than 70 percent.

2006

The Right Reverend *Katharine Jefferts Schori* is named the presiding bishop of the Episcopal Church. She is the first woman elected primate in the Anglican Communion.

2007

2007

Gail Kimbell is named Forest Service chief, the first woman in the job.

2007

2007

Historian *Drew Gilpin Faust* is named president of Harvard University, the first female in its almost four-hundred-year history.

Her Story

Bibliography

Adamson, Lynda G. *Notable Women in American History: A Guide to Recommended Biographies and Autobiographies.* Westport, Conn.: Greenwood Press, 1999.

Alvrez, Alicia. *The Ladies' Room Reader: The Ultimate Women's Trivia Book.* York Beach, Me.: Conari Press, 2000.

Bailey, Martha J. *American Women in Science.* Denver, Colo.: ABC-CLIO, 1994.

Barker-Benfield, G. J., and Catherine Clinton. *Portraits of American Women from Settlement to the Present.* New York: Oxford University Press, 1998.

Barnet, Andrea. *All-Night Party: The Women of Bohemian Greenwich Village and Harlem, 1913–1930.* Chapel Hill, N.C.: Algonquin Books, 2004.

Belford, Barbara. *Brilliant Bylines: A Biographical Anthology of Notable Newspaperwomen in America.* New York: Columbia University Press, 1986.

Bernikow, Louise. *The American Women's Almanac: An Inspiring and Irreverent Women's History.* New York: Berkley Books, 1997.

Brenner, Marie. *Great Dames: What I Learned from Older Women.* New York: Crown Publishers, 2000.

Cheney, Lynne. *A Is for Abigail: An Almanac of Amazing American Women.* New York: Simon and Schuster Books for Young Readers, 2003.

Colman, Penny. *Girls: A History of Growing Up Female in America.* New York: Scholastic, 2000.

——. *Rosie the Riveter: Women Working on the Home Front in World War II.* New York: Crown Publishers, 1995.

Collins, Gail. *America's Women: 400 Years of Dolls, Drudges, Helpmates and Heroines.* New York: HarperCollins, 2003.

Cooney, Miriam P., ed. *Celebrating Women in Mathematics and Science.* Reston, Va.: National Council of Teachers of Mathematics, 1996.

Cott, Nancy F., gen. ed. *The Young Oxford History of Women in the United States.* 11 vols. New York: Oxford University Press, 1995.

Cullen-DuPont, Kathryn. *The Encyclopedia of Women's History in America: Over 300 Years of Movements, Breakthroughs, Legislation, Court Cases, and Notable Women.* New York: Da Capo Press, 1996.

Drachman, Virginia G. *Enterprising Women: 250 Years of American Business.* Chapel Hill, N.C.: University of North Carolina Press, 2002.

Duncan, Jacci. *Making Waves: The 50 Greatest Women in Radio and Television.* Kansas City, Mo.: Andrews McMeel, 2001.

Epstein, Vivian Sheldon. *History of Women in Science for Young People.* Denver, Colo.: VSE, 1994.

Evans, Harold. *They Made America (From the Steam Engine to the Search Engine: Two Centuries of Innovators).* New York: Little, Brown, 2004.

Felder, Deborah G. *100 American Women Who Shaped American History.* San Mateo, Calif.: Bluewood Books, 2005.

——. *The 100 Most Influential Women of All Time: A Ranking Past and Present.* New York: Kensington, 2001.

——. *A Century of Women: The Most Influential Events in Twentieth-Century Women's History.* New York: Kensington, 1999.

——, Felder, Deborah G., and Diana Rosen. *Fifty Jewish Women Who Changed the World.* New York: Kensington, 2003.

Flexner, Eleanor, and Ellen Fitzpatrick. *Century of Struggle: The Woman's Rights Movement in the United States.* Cambridge, Mass.: Belknap Press, 1996.

Forbes, Malcolm, with Jeff Bloch. *Women Who Made a Difference.* New York: Simon and Schuster, 1990.

Garza, Hedda. *Barred from the Bar: A History of Women and the Legal Profession.* New York: Franklin Watts, 1996.

Gray, Gwendolyn. *Girls Who Grew Up Great: A Book of Encouragement for Girls About Amazing Women Who Dared to Dream.* Boulder, Colo.: Blue Mountain Press, 2003.

Griffiths, Sian, ed. *Beyond the Glass Ceiling: Forty Women Whose Ideas Shape the Modern World.* Manchester: Manchester University Press, 1996.

Harness, Cheryl. *Remember the Ladies: 100 Great American Women.* New York: HarperCollins, 2001.

Heinemann, Sue. *Timelines of American Women's History.* New York: Perigee Press, 1996.

History Makers. Bath, U.K.: Parragon, 1999.

Horwitz, Margot F. *A Female Focus: Great Women Photographers.* New York: Franklin Watts, 1996.

Hutchinson, Kay Bailey. *American Heroines: The Spirited Women Who Shaped Our Country.* New York: William Morrow, 2004.

Hymowitz, Carol, and Michaele Weissman. *A History of Women in America.* New York: Bantam Books, 1978.

Isaacs, Susan. *Brave Dames and Wimpettes: What Women Are Really Doing on Page and Screen.* New York: Ballantine, 1999.

Jaffe, Deborah. *Ingenious Women: From Tincture of Saffron to Flying Machines.* Phoenix Mill, U.K.: Sutton, 2003.

James, Edward T. *Notable American Women: A Biographical Dictionary.* 3 vols. Cambridge, Mass.: Belknap Press, 1971.

Jones, Constance. *1001 Things Everyone Should Know About Women's History.* New York: Doubleday, 2000.

Kass-Simon, G., and Patricia Farnes, eds. *Women of Science: Righting the Record.* Bloomington: Indiana University Press, 1990.

Kevles, Bettyann Holtzmann. *Almost Heaven: The Story of Women in Space.* New York: Basic Books, 2003.

King, Laurel. *Women of Power.* Berkeley, Calif.: Celestial Arts, 1989.

Ladies' Home Journal. *100 Most Important Women of the 20th Century.* Des Moines, Iowa: Ladies' Home Journal Books, 1998.

Lasher, Patricia, and Beverly Bentley. *Texas Women Interviews and Images.* Austin, Texas: Shoal Creek Publishers, 1980.

Leon, Vicki. *Uppity Women of the New World*. Berkeley, Calif.: Conari Press, 2001.

Lucey, Donna M. *I Dwell in Possibility–Women Build a Nation 1600–1920*. Washington, D.C.: National Geographic, 2001.

Lumme, Helena. *Great Women of Film*. New York: Billboard Books, 2002.

Lunardini, Christine. *What Every American Should Know About Women's History: 200 Events That Shaped Our Destiny*. Holbrook, Mass.: Adams Media, 1997.

Lyman, Darryl. *Great African-American Women*. New York: Gramercy Books, 1999.

Malone, John. *It Doesn't Take a Rocket Scientist: Great Amateurs of Science*. Hoboken, N.J.: John Wiley and Sons, 2002.

Marton, Kati. *Hidden Power: Presidential Marriages That Shaped Our Recent History*. New York: Pantheon Books, 2001.

McGinn, Elinor. *A Wide-Awake Woman: Josephine Roche in the Era of Reform*. Denver, Colo.: Colorado Historical Society, 2002.

McHenry, Robert, ed. *Famous American Women: A Biographical Dictionary from Colonial Times to the Present*. Mineola, N.Y.: Dover Publications, 1993.

Metz, Pamela, and Jacqueline Tobin. *The Tao of Women*. Atlanta: Humanics, 1995.

Morrow, Charlene, and Teri Perl, eds. *Notable Women in Mathematics: A Biographical Dictionary*. Westport, Conn.: Greenwood Press, 1998.

Movers and Shakers: The 100 Most Influential Figures in Modern Business. New York: Basic Books, 2003.

Munro, Eleanor. *Original: American Women Artists*. New York: Da Capo Press, 2000.

Ogilvie, Marilyn Bailey. *Women in Science: Antiquity Through the Nineteenth Century–A Biographical Dictionary with Annotated Bibliography*. Cambridge, Mass.: Massachusetts Institute of Technology Press, 1993.

O'Neill, Lois Decker. *The Women's Book of World Records and Achievements*. Garden City, N.Y.: Anchor Books, 1979.

Opdycke, Sandra. *The Routledge Historical Atlas of Women in America*. New York: Routledge, 2000.

Pierpont, Claudia Roth. *Passionate Minds: Women Rewriting the World*. New York: Vintage Books, 2000.

Portnow, Elaine. *The Quotable Woman: An Encyclopedia of Useful Quotations*. New York: Anchor Books, 1978.

Proffitt, Pamela, ed. *Notable Women Scientists*. Detroit: Gale Group, 1999.

Quinn, Tracy. *Quotable Women of the Twentieth Century*. New York: William Morrow, 1994.

Read, Phylllis J., and Bernard L. Witlieb. *The Book of Women's Firsts*. New York: Random House, 1992.

Reiter, Joan Swallow. *The Women: The Old West*. Alexandria, Va.: Time-Life Books, 1978.

Riley, Glenda, and Richard W. Etulain. *By Grit and Grace: Eleven Women Who Shaped the American West*. Golden, Colo.: Fulcrum Publishing, 1997.

Rimm, Sylvia. *How Jane Won: 55 Successful Women Share How They Grew from Ordinary Girls to Extraordinary Women*. New York: Crown Publishers, 2001.

Roehm, Michelle. *Girls Who Rocked the World 2: Heroines from Harriet Tubman to Mia Hamm*. Hillsboro, Ore.: Beyond Words Publishing, 2000.

Rose, Phyllis, ed. *The Norton Book of Women's Lives*. New York: W. W. Norton, 1993.

Schiff, Karenna Gore. *Lighting the Way: Nine Women Who Changed Modern America*. New York: Hyperion, 2005.

Seagraves, Anne. *Soiled Doves: Prostitution in the Early West*. Hayden, Id.: Wesanne Publications, 1994.

Shearer, Benjamin F., and Barbara S. Shearer. *Notable Women in the Life Sciences: A Biographical Dictionary*. Westport, Conn.: Greenwood Press, 1996.

——. *Notable Women in the Physical Sciences: A Biographical Dictionary*. Westport, Conn.: Greenwood Press, 1997.

Showell, Ellen H., and Fred M. B. Amram. *From Indian Corn to Outer Space: Women Invent in America*. Petersborough, N.H.: Cobblestone, 1995.

Sicherman, Barbara, and Carol Hurd Green. *Notable American Women: The Modern Period*. Cambridge, Mass.: Belknap Press, 1980.

Silver, A. David. *Enterprising Women: Lessons From 100 of the Greatest Entrepreneurs of Our Day*. New York: American Management Association, 1994.

Slater, Elinor, and Robert Slater. *Great Jewish Women*. Middle Village, N.Y.: Jonathan David Publishers, 1994.

Smith, Betsy Covington. *Breakthrough: Women in Law*. New York: Walker, 1984.

Smith, Lisa. *Nike Is a Goddess: The History of Women in Sports*. New York: Atlantic Monthly Press, 1998.

Sochen, June. *From Mae to Madonna: Women Entertainers in Twentieth-Century America*. Lexington: University Press of Kentucky, 1999.

Sochen, June. *Herstory: A Woman's View of American History*. New York: Alfred Publishing, 1974.

Sorel, Nancy Caldwell. *The Women Who Wrote the War*. New York: Arcade Publishing, 1999.

Stanley, Autumn. *Mothers and Daughter of Invention: Notes for a Revised History of Technology*. New Brunswick, N.J.: Rutgers University Press, 1993.

Stille, Darlene R. *Extraordinary Women Scientists*. Chicago: Children's Press, 1995.

Stratton, Joanna L. *Pioneer Women: Voices from the Kansas Frontier*. New York: Simon and Schuster, 1981.

Thimmeah, Catherine. *Girls Think of Everything: Stories of Ingenious Inventions by Women*. Boston: Houghton Mifflin, 2000.

Trager, James. *The Women's Chronology: A Year-by-Year Record, from Prehistory to the Present*. New York: Henry Holt, 1994.

Uglow, Jennifer S., comp. and ed. *The International Dictionary of Women's Biography*. New York: Continuum, 1985.

——. *The Northeastern Dictionary of Women's Biography*, 3rd ed. Boston.: Northeastern University Press, 1998.

Unforgettable Women of the Century. New York: People Books, 1998.

Vare, Ethlie Ann, and Greg Ptacek. *Mothers of Invention: From the Bra to the Bomb, Forgotten Women and their Unforgettable Ideas*. New York: Quill/William Morrow, 1987.

Varnell, Jeanne. *Women of Consequence: The Colorado Women's Hall of Fame*. Boulder, Colo.: Johnson Books, 1999.

Ventura, Varla. *Sheroes: Bold, Brash, and Absolutely Unabashed Superwomen from Susan B. Anthony to Xena*. Berkeley, Calif.: Conari Press, 1998.

Ware, Susan, ed. *Notable American Women: A Biographical Dictionary Completing the Twentieth Century*. Cambridge, Mass.: Belknap Press, 2004.

Webster's Dictionary of American Women. New York: Smithmark Publishers, 1996.

Welden, Amelie. *Girls Who Rocked the World: Heroines from Sacagawea to Sheryl Swoopes*. Hillsboro, Ore.: Beyond Words Publishing, 1998.

Williams, Pat and Ruth, with Michael Mink. *How to Be Like Women of Influence*. Deerfield Beach, Fla.: Health Communications, 2003.

Wilson, Vincent, Jr., and Gale S. McClung. *The Book of Distinguished American Women*. Brookeville, Md.: American History Research Associates, 2003.

Women Who Changed the World. London: Smith Davies, 2006.

Women Who Dared: A Book of Postcards. San Francisco: Pomegranate Artbooks, 1991.

Yentsch, Clarice M., and Carl J. Sindermann. *The Woman Scientist: Meeting the Challenges for a Successful Career*. New York: Plenum, 1992.

Yost, Edna. *American Women of Science*. New York: J. B. Lippincott, 1943.

——. *Women of Modern Science*. New York: Dodd, Mead, 1964.

Zahniser, J. D., comp. *And Then She Said . . . Quotations by Women for Every Occasion*. St. Paul, Minn.: Caillech Press, 1989.

Illustration Credits

Unless otherwise credited, book covers are courtesy of HarperCollins *Publishers*

Half title: Suffrage parade courtesy of the Library of Congress

Title page: Madeline Albright © AFP/Getty Images

Helen Keller courtesy of the Library of Congress

Woman in red dress print courtesy of the Library of Congress

Condoleezza Rice © Brooks Kraft/CORBIS

Page 8: Madeline Albright © Deborah Feingold/Corbis

Page 11: Alice Paul at women's rally courtesy of the Library of Congress

Oprah Winfrey © Mike Theiler/Reuters/Corbis

Page 13: Margaret Sanger courtesy of the Library of Congress

Page 15: Suffragist, "Mrs. Suffern," holding sign courtesy of the Library of Congress

1587: Baptism of Virginia Dare © Getty Images

1607: Engraving of Early Colonists at Jamestown, Virginia © Bettmann/CORBIS

1608: Pocahontas (1595-1617) illustration from "World Noted Women" by Mary Cowden Clarke, 1858 (engraving), Staal, Pierre Gustave Eugene (Gustave) (1817-82) (after)/Private Collection, Ken Welsh/The Bridgeman Art Library

1619: Dutch Ship Landing Slaves in Colonial America © Bettmann/CORBIS

1620: The Landing of the Pilgrims courtesy of the Library of Congress

1636: Harvard University courtesy of the Library of Congress

1637: Portrait of Anne Hutchinson (1591-1643), American School, (20th century)/Schlesinger Library, Radcliffe Institute, Harvard University,/The Bridgeman Art Library

1644:© Getty Images

1648 Margaret Brent © Louis Glanzman/National Geographic Image Collection

1649: Anne Bradstreet © The Granger Collection, New York

1660: Mary Dyer courtesy of the Library of Congress

1676: Kateri Tekakwitha © The Granger Collection, New York

1681: Penn's Treaty with the Indians, c.1830-40 (oil on canvas), Hicks, Edward (1780-1849)/Museum of Fine Arts, Houston, Texas, USA, Gift of Alice C. Simkins in memory of A. Nicholson Hanszen/The Bridgeman Art Library

1682: The Captivity of Mrs. Rowlandson courtesy of the Library of Congress

1692: Examination of a Witch, 1853 (oil on canvas), Matteson, Tompkins Harrison (1813-84)/© Peabody Essex Museum, Salem, Massachusetts, USA,/The Bridgeman Art Library

Tryals of Several Witches courtesy of the Library of Congress

Salem courtesy of the Library of Congress

1712: Hannah Callowhill Penn (1664-1726) c.1712 (oil on canvas), Hesselius, John (1728-78)/© Atwater Kent Museum of Philadelphia, Courtesy of Historical Society of Pennsylvania Collection,/The Bridgeman Art Library

1715: © Getty Images

1733: The earliest known view of Savannah, Georgia © The Granger Collection, New York

1744: Textile Dye Indigo © Getty Images

1758: Statue of Mary Jemison courtesy of the Library of Congress

1769: Portrait of Patience Wright © National Portrait Gallery, Smithsonian Institution / Art Resource, NY

1773: Mrs James Warren (Mercy Otis) c.1763 (oil on canvas), Copley, John Singleton (1738-1815)/Museum of Fine Arts, Boston, Massachusetts, USA, Bequest of Winslow Warren/The Bridgeman Art Library International

Phillis Wheatley Title page of *Poems on Various Subjects, Religious and Moral* by Phillis Wheatley, London: printed for A. Bell, 1773. Massachusetts Historical Society.Frontispiece of *Poems on Various Subjects, Religious and Moral* by Phillis Wheatley, London: printed for A. Bell, 1773. Massachusetts Historical Society

1775: Declaration of Independence, courtesy of the National Archives

1775: Battle of Lexington courtesy of the Library of Congress

1776: Shakers courtesy of the Library of Congress

Betsy Ross courtesy of the Library of Congress

Women at the polls in New Jersey © The Granger Collection, New York

Abigail Adams courtesy of the Library of Congress

1777: Sybil Ludington © The Granger Collection, New York

1778: Molly Pitcher courtesy of the Library of Congress

1780: Esther Reed © Getty Images

1781: Portrait of Elizabeth "Mumbet" Freeman (c.1742-1829) 1811 (w/c on ivory), Sedgwick, Susan Anne Livingston Ridley (fl.1811)/© Massachusetts Historical Society, Boston, MA, USA,/The Bridgeman Art Library International

1782: Deborah Sampson courtesy of the Library of Congress

1790: Susanna Rowson © Getty Images

Judith Sargent Stevens Murray © Terra Foundation for American Art, Chicago / Art Resource, NY

1793: Portrait of Catherine Ferguson used with Permission of Documenting the American South, the University of North Carolina at Chapel Hill Libraries

1796: Title page to "American Cookery," written by Amelia Simmons, published by Hudson & Goodwin, 1796 (print), American School, (18th century)/American Antiquarian Society, Worcester, Massachusetts, USA,/The Bridgeman Art Library International

1797: Elizabeth Ann Bayley Seton © The Granger Collection, New York

1805: Sacagawea with Lewis and Clark during their expedition of 1804-06 (color litho), Newell Convers Wyeth (1882-1945)/Private Collection, Peter Newark American Pictures/The Bridgeman Art Library International

1807: Cotton Gin © Getty Images

1809: Woman wearing hat courtesy of the Library of Congress

1813: The "Star Spangled Banner" courtesy of the Library of Congress

1814: Dolley Madison courtesy of the Library of Congress

1816: © The Granger Collection, New York

1819: Queen Kaahumanu by Madge Tennent, courtesy of The National Museum of Women in the Arts, Gift of Robert S. Rheem in honor of Constance Patterson Rheem.

1820: courtesy of the Library of Congress

1821: Emma Willard opens © Getty Images

1824: Portrait of Sarah Miriam Peale © National Portrait Gallery, Smithsonian Institution / Art Resource, NY

Margaret Bayard Smith undated oil on canvas by Charles Bird King, Gift of the Artist, courtesy of the Redwood Library and Athenaeum, Newport, Rhode Island

1825: Portrait of Rebecca Webb Lukens © Lukens National Historic Distric, Coatesville, PA

1827: Cherokee Nation Constitution courtesy of the Library of Congress

1829: Frances (Fanny) Wright courtesy of the Library of Congress

1831: The Liberator © The Granger Collection, New York

1832: Maria Weston Chapman © The Schlesinger Library, Radcliffe Institute, Harvard University

1833: Prudence Crandall © The Granger Collection, New York

Martha Coffin Wright courtesy of Smith College Collection, Smith College.

Lydia Maria Child © Getty Images

Audubon print © The Granger Collection, New York

Oberlin College courtesy of the Library of Congress

1834: "Lowell Offering," first published in 1841 as a "Repository of original articles written by Factory Girls' of Lowell, Massachusetts, and the first periodical in the world to be written exclusively by women © The Granger Collection, New York

1835: Paulina Kellogg Wright Davis courtesy of the Library of Congress

Harriot Kezia Hunt © The Schlesinger Library, Radcliffe Institute, Harvard University

1836: American missionary. Narcissa Whitman tending a sick Cayuse Indian at the Whitman Mission, about seven miles west of present-day Walla Walla, Washington: line engraving, 19th century © The Granger Collection, New York

1837: Sarah Josepha Hale courtesy of the Library of Congress

Mary Lyon courtesy of the Library of Congress

1838: Angelina Grimké © The Granger Collection, New York

Abby Kelley Foster © The Schlesinger Library, Radcliffe Institute, Harvard University

Rebecca Gratz courtesy of the Library of Congress

1839: Margaret Fuller courtesy of the Library of Congress

Sarah Grimké courtesy of the Library of Congress

1840: Anti-Slavery Society, including Lucretia Mott (1793-1880), American Photographer, (19th century)/Schlesinger Library, Radcliffe Institute, Harvard University,/The Bridgeman Art Library International

Lucretia Coffin Mott contemporary American engraving © The Granger Collection, New York

Margaret Haughery © Robert Holmes/CORBIS

Ernestine Rose courtesy of the Library of Congress

1843: Dorothea Dix courtesy of the Library of Congress

Sojourner Truth courtesy of the Library of Congress

1844: "Hymn for the Working Children" by Fanny J. Crosby courtesy of the Library of Congress

Portrait of Fanny Crosby © The Granger Collection, New York

The factories of Lowell, Massachusetts, on the banks of the Merrimac River: American engraving, 1844 © The Granger Collection, New York

1848: Elizabeth Cady Stanton courtesy of the Library of Congress

Portrait of Maria Mitchell © The Granger Collection, New York

1849: Lithographs by Frances F. Palmer courtesy of the Library of Congress

Portrait of Elizabeth Blackwell © The Granger Collection, New York

1850: Female medical students dissect cadavers during anatomy class at the Women's Medical College of Pennsylvania, located in Philadelphia. © Bettmann/CORBIS

Lucy Stone Banner © Mary Evans/the Women's Library

Portrait of Lucy Stone courtesy of the Library of Congress

Harriet Tubman photographs courtesy of the Library of Congress

1851: Amelia Bloomer © The Granger Collection, New York

Myrtilla Miner courtesy of the Library of Congress

1852: Emily Dickinson © Getty Images

Catharine Beecher © The Granger Collection, New York

Matilda Joslyn Gage courtesy of the Library of Congress

Camilla Urso © Getty Images

Harriet Beecher Stowe courtesy of the Library of Congress

1853: Antoinette Brown Blackwell courtesy of the Library of Congress

Mary Ann Shadd Cary © The Granger Collection, New York

Harriet Goodhue Hosmer courtesy of the Library of Congress

1855: Elizabeth Keckley © Getty Images

Emeline Horton Cleveland courtesy of Drexel University College of Medicine; Archives and Special Collections on Women in Medicine

Printing office © The Granger Collection, New York

1857: Emily Howland courtesy of The Cayuga County Historian's Office

1860: Frances E. W. Harper courtesy of the Library of Congress

Ellen Curtis Demorest © Getty Images

Ellen White © Ellen G. White Estate, Inc.

Chinese Slave Girl, © 1905, Postcard, Collection of the Oakland Museum of California

Elizabeth Peabody courtesy of the Library of Congress

Mary Edwards Walker courtesy of the Library of Congress

1861: Union soldiers marching through a city street on their way to join the Civil War courtesy of the Library of Congress

Belle Boyd courtesy of the Library of Congress

Sally Tompkins courtesy of the Library of Congress

1862: Mary Bickerdyke courtesy of the Library of Congress

Marie Elizabeth Zakrzewska courtesy of The Schlesinger Library, Radcliffe Institute, Harvard University

Mary Jane Patterson courtesy of Oberlin College Archives, Oberlin, Ohio

Julia Ward Howe courtesy of the Library of Congress

1864: Line drawing from Thirteenth Annual Announcement of The New-England Female Medical College, original item held at the Harvard Medical Library in the Francis A. Countway Library of Medicine

Portrait of Mary Rice Livermore © CORBIS

1865: Pauline Cushman courtesy of the Library of Congress

Hans Brinker © The Granger Collection, New York

1866: Lucy Taylor courtesy of the Kansas State Historical Society

Vinnie Ream Hoxie courtesy of the Library of Congress

1867: Edmonia Lewis © The Granger Collection, New York

1868: Little Women Cover © Mary Evans Picture Library

Myra Bradwell ©Mary Evans/the Women's Library

Little Girl in a Blue Armchair, 1878 (oil on canvas) (signature cropped), Cassatt, Mary Stevenson (1844-1926)/Mellon Coll., Nat. Gallery of Art, Washington DC, USA./The Bridgeman Art Library International

Portrait Mary Cassatt courtesy of the Library of Congress

1869: Susan B. Anthony courtesy of the Library of Congress

Fanny Jackson Coppin courtesy of the Library of Congress

Arabella Mansfield courtesy of DePauw University Archives and Special Collections

Transcontinental railroad illustration courtesy of the Library of Congress

1870: Photograph of woman in corset courtesy of the Library of Congress

Shop display corset with slogan, "The Celebrated CB Corset', 19th century (photo)/Killerton, Devon, UK, National Trust Photographic Library/Andreas von Einsiedel/The Bridgeman Art Library International

Women working courtesy of the Library of Congress

Calamity Jane courtesy of the Library of Congress

1871: Sophia Smith courtesy of Smith College Archives, Smith College Rockwood Photographer, 839 Broadway, New York, undated portrait

1872: Victoria Woodhull © Bettmann/CORBIS

Polly Bemis courtesy Asian American Comparative Collection, University of Idaho, Moscow, and Johnny and Pearl Carrey, www.uidaho.edu/LS/AACC/

Mary Putnam Jacobi courtesy of the Library of Congress

1873: Censorship seal © The Granger Collection, New York

1874: Lawn Tennis © The Granger Collection, New York

The Women's Christian Temperance Union (WCTU) © Getty Images

Frances Willard courtesy of the Library of Congress

1875: Virginia Minor © Getty Images

Packaging of Lydia Pinkham's Vegetable Tonic, c.1940 (b/w photo), American Photographer, (20th century)/Schlesinger Library, Radcliffe Institute, Harvard University/The Bridgeman Art Library International

1876: Martha Maxwell courtesy of Boulder Historical Society Collection of the Carnegie Branch Library for Local History

1878: Clara Shortridge Foltz courtesy of the Bancroft Library, University of California, Berkeley

1879: Mary Baker Eddy courtesy of the Library of Congress

Belva Lockwood courtesy of the Library of Congress

Ida Lewis courtesy of the Library of Congress

1880: Anna Howard Shaw courtesy of the Library of Congress

Susette LaFlesche Tibbles courtesy of the Nebraska Historical Society #RG2026-PH

Women Delegates of Knights of Labor © Bettmann/CORBIS

1881: Clara Barton courtesy of the Library of Congress

1882: Elizabeth Agassiz © The Granger Collection, New York

Christine Ladd-Franklin courtesy of the Library of Congress

Ellen Swallow Richards courtesy of the Library of Congress

Minnie Maddern Fiske photograph and poster courtesy of the Library of Congress

1883: Sarah Winnemucca © The Granger Collection, New York

Banner of the Civil Service Women's Suffrage Society © Mary Evans/the Women's Library

Harvey's Restaurant courtesy of the Library of Congress

Emma Lazarus © Getty Images

Statue of Liberty courtesy of the Library of Congress

1884: Sita and Sarita, or Young Girl with a Cat, 1893-94 (oil on canvas), Beaux, Cecilia (1855-1942)/Musee d'Orsay, Paris, France, Lauros / Giraudon/The Bridgeman Art Library International

Becker sisters branding cattle courtesy Colorado Historical Society (F28537) All Rights Reserved

1885: Maud Powell courtesy of the Library of Congress

Annie Oakley © The Granger Collection, New York

1886: Grace Hoadley Dodge courtesy of University Archives, Columbia Universiry in the City of New York

A rendering of the Italian Renaissance style Women's Building designed by Sophia G. Hayden, the first woman awarded a degree in architecture from MIT © Getty Images

Mrs. Lucy E. Parsons photographed by Chicago Daily News, courtesy of Chicago History Museum DN-0063954

1887: Anne Sullivan Macy courtesy of the Library of Congress

Frances Wisebart Jacobs window courtesy of the Denver Public Library, Western History Collection, Elaine A. Clearfield 920.0788 C580u

1888: Mother Marianne Cope Courtesy of the Sisters of St. Francis, Syracuse, NY

Sissieretta Jones courtesy of the Library of Congress

1889: Self-Portrait by Lilla Cabot Perry © Terra Foundation for American Art, Chicago / Art Resource, NY

Anna Bissell compliments of BISSELL Homecare, Inc.

Jane Addams courtesy of the Library of Congress

Nellie Bly courtesy of the Library of Congress

Kate Chopin © The Granger Collection, New York

Susan La Flesche Picotte © The Granger Collection, New York

Frances Benjamin Johnston courtesy of the Library of Congress

Jane Cunningham Croly courtesy of Sophia Smith Collection, Smith College, Fanny Fern and Ethel Parton Papers

Mother Frances Xavier Cabrini courtesy of the Library of Congress

Emma Goldman courtesy of the Library of Congress

1890:Lucy Craft Laney courtesy of Georgia Capitol Museum, Office of Secretary of State

Daughters of the American Revolution courtesy of the Library of Congress

Alice Stone Blackwell courtesy of the Library of Congress

Kate Gleason © Bettmann/CORBIS

Ida Gray Nelson courtesy of Manuscripts, Archives and Rare Books Division, Schomburg Center for Research in Black Culture, the New York Public Library, Astor, Lenox and Tilden Foundations

1891: Sister Mary Katharine Drexel provided by the Archives of the Sisters of the Blessed Sacrament

Harriet Maxwell Converse courtesy of the New York State Library

Promotional Portrait of Martha Matilda Harper, courtesy of Jane R. Plitt, Harper biographer

Mary Emma Woolley courtesy of the Library of Congress

1892: Anna Julia Cooper © The Granger Collection, New York

Charlotte Perkins Gilman courtesy of the Library of Congress

Bertha Palmer courtesy of the Library of Congress

Ellis Island courtesy of the Library of Congress

1893: Hannah G. Solomon courtesy of the Library of Congress

Florence Bascom courtesy of Sophia Smith Collection, Smith College (Florence Bascom Papers)

Annie Laurie courtesy of the Library of Congress

1894: Josephine St. Pierre Ruffin used with permission of documenting the American South, The University of North Carolina at Chapel Hill Libraries

Ethel Barrymore courtesy of the Library of Congress

1895: Katherine Lee Bates © The Granger Collection, New York

Ida B. Wells-Barnett © The Granger Collection, New York

Catherine Furbish courtesy of Bowdoin College Library

Catherine Evans Whitener courtesy of the Whitfield-Murray Historical Society

Mary Engle Pennington courtesy of the Collections of the University of Pennsylvania Archives

Nampeyo courtesy of the Library of Congress

Lillian D. Wald © The Granger Collection, New York

1896: Amy Marcy Cheney Beach courtesy of the Library of Congress

The Queen's Twin cover courtesy of the Library of Congress

Fannie Farmer © The Granger Collection, New York

Mary Church Terrell courtesy of the Library of Congress

Annie Jump Cannon courtesy of the Library of Congress

1897: Alice McLellan Birney courtesy of the Library of Congress

1898: Julia Morgan © The Granger Collection, New York

1899: Florence Kelley courtesy of the Library of Congress

Rosa Minoka-Hill © Getty Images

1900: Margaret Abbott courtesy of the Chicago Tribune

Child worker courtesy of the Library of Congress

Women workers courtesy of the Library of Congress

Nannie Helen Burroughs courtesy of the Library of Congress

Carrie Chapman Catt courtesy of the Library of Congress

1901: Zitkala Sa courtesy of the Library of Congress

Jessie Field Shambaugh photo courtesy of Iowa State University Extension and Ruth Shambaugh Watkins, daughter of Jessie Field Shambaugh

Sophonisba Breckinridge courtesy of the Library of Congress

Maud Wood Park courtesy of the Library of Congress

1902: Alice Cunningham Fletcher © Getty Images

Annie Malone courtesy of Annie Malone Children & Family Service Center

Harriet Stanton Blatch courtesy of the Library of Congress

Agnes Nestor courtesy of the Library of Congress

1903: Molly Picon courtesy of the Library of Congress

Mother Jones courtesy of the Library of Congress

Rose Schneiderman courtesy of the Library of Congress

Maggie Lena Walker © The Granger Collection, New York

Helen Keller courtesy of the Library of Congress

1904: Mary McLeod Bethune courtesy of the Library of Congress

Lena Bryant courtesy of the Library of Congress

Ida Tarbell courtesy of the Library of Congress

1905: Mary Chesnut © The Granger Collection, New York

Nettie Stevens with permission of the University Archives, Columbia University in the City of New York

Madam C. J. Walker © The Granger Collection, New York

1906: Fanny Bullock Workman courtesy of the Library of Congress

Williamina Stevens Fleming courtesy of Harvard University Archives, call Nº HUV 1210 (9-6)

Sophie Tucker courtesy of the Library of Congress

1907: Japanese Brides Lining up for Inspection © Bettmann/CORBIS

1908: Edith Abbott courtesy of Special Collections Research Center, University of Chicago Library

1909: Mary White Ovington © Bettmann/CORBIS

Helen Hayes courtesy of the Library of Congress

Gertrude Stein courtesy of the Library of Congress

Alva Erskine Smith Vanderbilt Belmont courtesy of the Library of Congress

1910: Charlotte Vetter Gulick courtesy of the Library of Congress

Crystal Eastman courtesy of the Library of Congress

Beautician and businesswoman Elizabeth Arden with a group of nurses. © Getty Images

Advertisement in *Vogue*, 28 April 1937 © Mary Evans Picture Library

1911: Annie Smith Peck courtesy of the Library of Congress

Mae West © Getty Images

Harriet Quimby courtesy of the Library of Congress

Students practice typing drills in class at the Katharine Gibbs School © Associated Press

Triangle Shirtwaist Company factory courtesy of the Library of Congress

Maud Slye courtesy of Special Collections Department, W.E.B. DuBois Library, University of Massachusetts, Amherst

Gracie Allen courtesy of the Library of Congress

1912: Mary Antin © Chicago History Museum

Elizabeth Gurley Flynn courtesy of the Library of Congress

Julia Lathrop courtesy of the Library of Congress

Henrietta Swan Leavitt © The Granger Collection, New York

Josephine Roche courtesy of the Library of Congress

Anita Loos courtesy of the Library of Congress

Lucy Burns courtesy of the Library of Congress

Alice Paul courtesy of the Library of Congress

Abigail Scott Duniway courtesy of the Library of Congress

Henrietta Szold courtesy of the Library of Congress

Juliette Gordon Low © Bettmann/CORBIS

1913: Belle Moskowitz courtesy of Elisabeth Israels Perry

Elsie de Wolfe courtesy of the Library of Congress

1914: Helena Rubinstein © Getty Images

Advertisement for beauty products of Helena Rubinstein © Mary Evans Picture Library

Mother and child courtesy of the Library of Congress

1915: Marianne Moore courtesy of the Library of Congress

Ruth St. Denis courtesy of the Library of Congress

Hazel Harrison courtesy of the Ohio Historical Society

The Women's Peace Party courtesy of the Library of Congress

Ruth Sawyer photograph from Ruth Sawyer Collection, College of St. Catherine Archives

1916: Margaret Sanger photographs courtesy of the Library of Congress

Mary Pickford courtesy of the Library of Congress

Hetty Green courtesy of the Library of Congress

Pansy, 1926 (oil on canvas), O'Keeffe, Georgia (1887-1986)/Brooklyn Museum of Art, New York, USA, © DACS/The Bridgeman Art Library International

Georgia O'Keeffe (1887-1986) 1920 (silver gelatin print), Stieglitz, Alfred (1864-1946)/Private Collection, Archives Charmet/The Bridgeman Art Library International

1934: Tampax advertisement courtesy of Procter & Gamble Company

Ella Fitzgerald courtesy of the Library of Congress

Mahalia Jackson courtesy of the Library of Congress, Prints & Photographs Division, Carl Van Vechten Collection, (reproduction number, e.g., LC-USZ62-54234)

Mary Margaret McBride courtesy of the Library of Congress

Ruth Benedict © The Granger Collection, New York

Rosalind Russell © Getty Images

1935: Muriel Rukeyser courtesy of the Library of Congress

Mari Sandoz courtesy of the Library of Congress

Effa Manley © Associated Press

Sadie Alexander © Bettmann/CORBIS

Ginger Rogers © AFP/Getty Images

Shirley Temple © AFP/Getty Images

Billie Holiday courtesy of the Library of Congress

1936: Dorothy Kilgallen © Bettmann/CORBIS

Margaret Mitchell courtesy of the Library of Congress

Margaret Rudkin courtesy of the Library of Congress

Helen Stephens © Bettmann/CORBIS

Margaret Bourke-White courtesy of the Library of Congress

Photograph by Dorothea Lange courtesy of the Library of Congress

Dorothy Thompson courtesy of the Library of Congress

1937: Rose Blumkin © Associated Press

Hattie Alexander © Bettmann/CORBIS

1938: Emma Tenayuca UTSA's Institute of Texan Cultures, San Antonio Light Collection, №L-1544-D, Courtesy of the Hearst Corporation

Marjorie Kinnan Rawlings courtesy of the Library of Congress

Catherine Stern, Ph.D. courtesy of Fred Stern

Katherine Blodgett courtesy of the Library of Congress

Photograph by Marion Post Wolcott courtesy of the Library of Congress

Pearl S. Buck courtesy of the Library of Congress

1939: Karen Horney © Bettmann/CORBIS

Elsie Clews Parsons courtesy of the American Philosophical Society

Thérèse Bonney courtesy of the Library of Congress

Berenice Abbott © Allen Ginsberg/CORBIS

Photograph by Berenice Abbott courtesy of the Library of Congress

Anna Lee Aldred © Associated Press

The Carter Family © Michael Ochs Archives/Corbis

Marian Anderson © Associated Press

Zora Neale Hurston courtesy of the Library of Congress

Grandma Moses © Getty Images

1940: Judy Garland courtesy of the Library of Congress

May Hill Arbuthnot © 2003 Grosset and Dunlap, *Storybook Treasury of Dick and Jane and Friends*

Mirror Image 1, 1969 (painted wood), Nevelson, Louise (1900-88)/Museum of Fine Arts, Houston, Texas, USA. © DACS /Gift of The Brown Foundation, Inc./The Bridgeman Art Library International

Minnie Pearl © Getty Images

Dale Messick © Associated Press

Hattie McDaniel © Getty Images

Margaret Chase Smith © Getty Images

1941: Wonder Woman #21 © 1974 DC Comics. All Rights Reserved. Used with Permission.

Florence Seibert © Bettmann/CORBIS

December 7 poster courtesy of the Library of Congress

Hilda Terry © Daily News

Essie Parrish Courtesy of the Phoebe Apperson Hearst Museum of Anthropology and the Regents of the University of California-photographed by Samuel Barrett 25-5798

1942: Sarah Vaughan courtesy of the Library of Congress

Elizabeth Taylor © Getty Images

Jackie Cochran © Getty Images

Agnes de Mille courtesy of the Library of Congress

Oveta Culp Hobby courtesy of the Library of Congress

Sylvia Porter courtesy of the Library of Congress

1943: Lena Horne © Time & Life Pictures/Getty Images

Constance Baker Motley courtesy of the Library of Congress

Ayn Rand courtesy of the Library of Congress

Georgette "Dicky" Chappelle courtesy of the Library of Congress

Euphemia Lofton Haynes courtesy of Haynes-Lofton Family Papers, The American Catholic History Research Center and University Archives, The Catholic University of America Washington, DC

Grace Murray Hopper courtesy of the Library of Congress

"Rosie the Riveter" photograph courtesy of the Library of Congress

1944: Helen Brooke Taussig courtesy of the Library of Congress

Pauli Murray courtesy of the Library of Congress

Chien-Shiung Wu courtesy of the Library of Congress

Angela Lansbury courtesy of the Library of Congress

Lauren Bacall © Getty Images

War veterans and coeds taking notes during classroom lecture at crowded University of Iowa where there are now 6,000 vets, which is 60% of the school's enrollment due to financing by the GI Bill of Rights © Time & Life Pictures/Getty Images

1945: Joan Crawford courtesy of the Library of Congress

Ruth Bigelow courtesy of Cindi Bigelow, President at Bigelow Tea

The first twelve women enter Harvard Medical School courtesy of Harvard Medical Library collection at the Count-way Library of Medicine

Woman worker courtesy of the Library of Congress

1946: Emily Greene Balch courtesy of the Library of Congress

Mary Ritter Beard courtesy of the Library of Congress

Jayne Meadows © John Springer Collection/CORBIS

Edith Houghton courtesy of National Baseball Hall of Fame Library, Cooperstown, N.Y.

Estée Lauder courtesy of the Library of Congress

Marilyn Monroe © Getty Images

1947: Dorothy Stimson Bullitt courtesy of University of Washington's Special Collections Division

Marjory Stoneman Douglas © Morton Beebe/CORBIS

Gerty Radnitz Cori © The Granger Collection, New York

Alice Hamilton courtesy of the Library of Congress

Jessica Tandy courtesy of the Library of Congress

Celia Cruz courtesy of the Library of Congress

Edith Clarke © Bettmann/CORBIS

1948: Alice Coachman courtesy of the Library of Congress

Mary Agnes Hallaren courtesy of the Library of Congress

1949: Stella Adler © Getty Images

Yoshiko Uchida courtesy of The Bancroft Library, University of California, Berkeley, 1986. 0591:259

Levittown Housing Development © Time & Life Pictures/Getty Images

Gwendolyn Brooks courtesy of the Library of Congress

Mary Martin courtesy of the Library of Congress

Georgia Neese Clark courtesy of the Library of Congress

Gertrude Berg © Time & Life Pictures/Getty Images

1950: Roberta Peters © Getty Images

Malvina Reynolds © Ted Streshinsky/CORBIS

Olive Ann Beech © Bettmann/CORBIS

Rachel Fuller Brown and Elizabeth Lee Hazen courtesy of Photography and Illustration Wadsworth Center New York State Department of Health

Hisaye Yamamoto courtesy of the Bancroft Library, University of California, Berkeley

Helen Frankenthaler © Getty Images

Beatrice Hicks © Getty Images

Babe Didrikson Zaharias courtesy of the Library of Congress

Althea Gibson playing tennis courtesy of the Library of Congress

Althea Gibson with trophy © Bettmann/CORBIS

Jade Snow Wong © 1989 University of Washington Press

1951: Shelley Winters © Getty Images

Lee Krasner © Arnold Newman/Getty Images

Lucille Ball courtesy of the Library of Congress

Maggie Higgins © Time & Life Pictures/Getty Images

Lillian Vernon © Associated Press

Brownie Wise courtesy of Tupperware

Bette Nesmith Graham from "Letter Perfect", newsletter of the Liquid Paper Corporation, Dallas, TX. Special Commemorative Issue, November 1975, pp. 1-2. Newsletter from the Collections at the University of North Texas Archives. Reproduced with permission from Liquid Paper Corporation.

Hannah Arendt courtesy of the Library of Congress

1952: Leontyne Price courtesy of the Library of Congress

Lillian Hellman courtesy of the Library of Congress

Frances Horwich © Associated Press

Flannery O'Connor courtesy of Flannery O'Connor Collection, Georgia College & State University Library

Doriot Anthony Dwyer courtesy of Boston University, College of Fine Arts

Gertrude Dunn Baseball card courtesy of Larry Fritsch Cards, Inc. P.O. Box 863 Stevens Point , WI. 54481

Barbara Holdridge and Marianne Mantell courtesy of the Caedmon archives

Child's Christmas in Wales album cover courtesy of the Caedmon archives

Virginia Apgar courtesy of the Library of Congress

1953: Mary Steichen Calderone courtesy of the Library of Congress

Clare Boothe Luce courtesy of the Library of Congress

Betty White © Getty Images

Lupe Serrano as Odile and Royes Fernandez as Prince Siegfried in Swan Lake, III PHOTO: Jack Mitchell

1954: Maria Tallchief courtesy of the Library of Congress

1955: Beverly Sills courtesy of the Library of Congress

Dinah Shore © Associated Press

Edith Green courtesy of the Library of Congress

Rosa Parks on bus courtesy of the Library of Congress

Rosa Parks is fingerprinted © Associated Press

Shirley MacLaine © CinemaPhoto/Corbis

1956: La Leche League International founders (1956) courtesy of La Leche League International

Tina Turner © Getty Images

Josephine Perfect Bay courtesy of The Bay and Paul Foundation Archive

Rose Hum Lee courtesy of Department of Special Collections, Charles E. Young Research Library, UCLA

Margaret Towner © CORBIS

1957: Patsy Cline © Getty Images

Ella Baker courtesy of the Library of Congress

Sadie Ginsberg courtesy of Judith Ginsberg Bender

Alice Herrington courtesy of Friends of Animals

Sister Rose Thering courtesy of Seton Hall

Chita Rivera courtesy of the Library of Congress

1958: Ethel Percy Andrus © Bettmann/CORBIS

Joyce Chen courtesy of the family of Joyce Chen

1959: Carol Burnett © Getty Images

Lorraine Hansberry courtesy of the Library of Congress

Ruth Handler © Getty Images

1960: Joan Baez courtesy of the Library of Congress

Suzanne Farrell courtesy of the Library of Congress

Jerrie Cobb courtesy of the Library of Congress

Harper Lee photograph courtesy of the Library of Congress

Sadie Neakok photograph by Edwin Hall, Jr. Used by permission of the University of Washington Press

Wilma Rudolph courtesy of the Library of Congress

Nancy Dickerson Whitehead courtesy of the Library of Congress

Irmgard Flugge-Lotz courtesy of Stanford University Archives

Birth control pill © Time & Life Pictures/Getty Images

1961: Jane Jacobs © Getty Images

Private school for retarded children visited by Mrs. Sargent R. Shriver. © Time & Life Pictures/Getty Images

Esther Peterson © Getty Images

Tillie Olsen © Christopher Felver/CORBIS

1962: Rachel Carson © Getty Images

Katherine Anne Porter courtesy of the Library of Congress

Felice Schwartz © CORBIS

Barbara Tuchman © Associated Press

Dolores Huerta © Getty Images

Barbra Streisand courtesy of the Library of Congress

Dr. Frances Kelsey courtesy of the Library of Congress

Rita Moreno courtesy of the Library of Congress

Jacqueline Kennedy Onassis © Associated Press

1963: Mary Kay Ash © Getty Images

Maria Goeppert-Mayer courtesy of the Library of Congress

Jean Nidetch © Getty Images

Annie Dodge Wauneka © Getty Images

Betty Friedan Speaking at Political Rally in New York © JP Laffont/Sygma/CORBIS

Portrait of Betty Friedan © Associated Press

Elaine De Kooning © Arnold Newman/Getty Images

Equal rights buttons © David J. & Janice L. Frent Collection/Corbis

Julia Child © Arnold Newman/Getty Images

1964: Patricia Neal courtesy of the Library of Congress

Yoko Ono © Getty Images

Kitty Genovese Crime Scene © Time & Life Pictures/Getty Images

Fannie Lou Hamer courtesy of the Library of Congress

Virginia Satir courtesy of Department of Special Collections, Davidson Library, University of California, Santa Barbara

Civil rights courtesy of the Library of Congress

Carol Channing © Getty Images

Ruth Duckworth © Associated Press

1965: Twyla Tharp © Eve Arnold/Magnum Photos

Factor VIII © Vo Trung Dung/corbis Sygma

Ruth Fertel courtesy of Ruth's Chris Steak House

Stephanie Kwolek photograph by Michael Branscom. Courtesy of the Lemelson-MIT Program

Lady Bird Johnson © Getty Images

Shirley Muldowney © Getty Images

1966: Barbara Jordan courtesy of the Library of Congress

The National Organization for Women (NOW) © Bettmann/CORBIS

Marlo Thomas © Getty Images

Yvonne Brathwaite-Burke courtesy of the Library of Congress

Susan Sontag courtesy of the Library of Congress

Eve Queler courtesy of Eve Queler

Wilma Vaught © AFP/Getty Images

Betsy Ancker-Johnson AIP Emilio Segre Visual Archives, Physics Today Collection

1967: Muriel Siebert © Bettmann/CORBIS

Dian Fossey © Yann Arthus-Bertrand/CORBIS

K. Switzer © Associated Press

Ida Rolf photograph used with permission from the Rolf Institute® of Structural Integration

Mary Sinclair © Regents of the University of Michigan

Helen Claytor © Associated Press

1968: Erma Bombeck © Douglas Kirkland/CORBIS

Joan Didion © Associated Press

Goldie Hawn © Getty Images

Shirley Chisholm courtesy of the Library of Congress

Diahann Carroll © Getty Images

Mary Washington Wylie courtesy of Barbara Shepherd

Peggy Fleming © Time & Life Pictures/Getty Images

Miss America protest © Associated Press

1969: Marilyn Horne © Time & Life Pictures/Getty Images

Judi Sheppard Missett courtesy of Jazzercise, Inc.

Tania León © 1994 New World Records/CRI

Joan Ganz Cooney © Time & Life Pictures/Getty Images

Jane Fonda © Getty Images

Elisabeth Kübler-Ross © Bettmann/CORBIS

NARAL © Time & Life Pictures/Getty Images

Coretta Scott King courtesy of the Library of Congress

Jessica McClintock © Roger Ressmeyer/CORBIS

Golda Meir courtesy of the Library of Congress

1970: Pearl Bailey ©2004 Living Era

Annie Leibovitz © Getty Images

Arlene Blum © Associated Press

Kate Millett © Sophie Bassouls/CORBIS SYGMA

Eleanor Holmes Norton © Associated Press

Ruth Westheimer © Getty Images

LaDonna Harris © Bettmann/CORBIS

First edition of *Our Bodies, Ourselves* courtesy of Our Bodies Ourselves. The 8th edition of *Our Bodies, Ourselves* was published in 2005

Diane Crump © Time & Life Pictures/Getty Images

Ada Louise Huxtable © Getty Images

Mary Tyler Moore © Getty Images

Vera Rubin photograph courtesy of Mark Godfrey, Director of Photography, The Nature Conservancy

1971: Carole King © Getty Images

Helen Caldicott © Time & Life Pictures/Getty Images

Bella Abzug courtesy of the Library of Congress

Letty Cottin Pogrebin © Time & Life Pictures/Getty Images

Gloria Steinem courtesy of the Library of Congress

Jeanne M. Holm © Bettmann/CORBIS

Maggie Kuhn © Associated Press

1972: Bette Midler © Getty Images

Maya Angelou © Getty Images

Linda Ellerbee © Associated Press

Eva Hesse © Time & Life Pictures/Getty Images

Patricia Schroeder © Associated Press

Susan Stamberg © Time & Life Pictures/Getty Images

Eudora Welty courtesy of the Library of Congress

Diane Arbus © Getty Images

Gertrude Boyle tells Microsoft Corp. Chairman Bill Gates that after all of the years that her company has used his products, that he needs to try one of her company's jackets © Associated Press

Juanita Kreps © Time & Life Pictures/Getty Images

Sally J. Priesand © Associated Press

Title IX © Associated Press

ERA Supporter © Getty Images

1973: Barbara Stanwyck courtesy of the Library of Congress

Marian Wright Edelman © Getty Images

Shirley Ann Jackson © Associated Press

Billie Jean King © Getty Images

Julia Phillips © Getty Images

Adrienne Rich courtesy of the Library of Congress

Sentinel by Anne Truitt © Albright-Knox Art Gallery/CORBIS

Sarah Weddington © Time & Life Pictures/Getty Images

1974: Andrea Dworkin © Time & Life Pictures/Getty Images

Karen Silkwood © Mark Peterson/Corbis

Two of the 15 women (front row) who entered the U.S. Merchant Marine Academy in 1974 assemble with their platoon to begin a two-week military indoctrination period in July of that year. Photograph courtesy of U.S. Merchant Marine Academy

Katharine Meyer Graham © Doug Menuez/CORBIS

Betty Ford photographs courtesy of the Library of Congress

Mary McGrory © Time & Life Pictures/Getty Images

Joyce Meskis courtesy of Joyce Meskis

Helen Thomas courtesy of the Library of Congress

Chris Evert © Getty Images

1975: Dolly Parton © Time & Life Pictures/Getty Images

Gilda Radner © Getty Images

Amy Alcott courtesy of Amy Alcott

Susan Brownmiller © Time & Life Pictures/Getty Images

Jacket cover from *The War Against The Jews: 1933-1945* by Lucy S. Dawidowicz. Used by permission of Bantam Books, a division of Random House, Inc.

Joan Rivers © Getty Images

Carole Little © AFP/Getty Images

Martina Navratilova © AFP/Getty Images

Ella Grasso © Getty Images

Virginia Hamilton © 2008. Used by permission of Arnold Adoff

1976: Maxine Hong Kingston © Time & Life Pictures/Getty Images

Oneida Bingo and Casino © Associated Press

Linda Alvarado photograph courtesy of Alvarado Construction, Inc.

Elizabeth Claiborne Ortenberg © Getty Images

Jane Pauley © Getty Images

Julia Robinson courtesy of George Bergman

Barbara Walters © Getty Images

Sarah Caldwell © Time & Life Pictures/Getty Images

Dixy Lee Ray © Time & Life Pictures/Getty Images

1977: Janet Guthrie © Getty Images

Kay Koplovitz © Getty Images

Leslie Silko © Christopher Felver/Corbis

Rosalyn Yalow courtesy of the Library of Congress

National Women's Conference © Bettmann/CORBIS

Debbi Fields © Associated Press

Mi Casa Resource Center for Women courtesy of Juana Bordes

Josie Cruz Natori © Andreea Angelescu/Corbis

Ann Fudge © Getty Images

Meryl Streep © Getty Images

Lily Tomlin © Getty Images

1978: Nancy Landon Kassebaum © Time & Life Pictures/Getty Images

Patricia Locke courtesy of Kevin Locke

Judith Resnik © Getty Images

Faye Wattleton © Bettmann/CORBIS

Vivian Scott of Melcher, Iowa points to a 1942 photograph of herself in her Women's Army Auxiliary Corps uniform © Associated Press

1979: Octavia Butler © Beth Gwinn

Judy Chicago © Time & Life Pictures/Getty Images

Detail of "The Dinner Party" © AFP/Getty Images

Sexual Harassment of Working Women cover © 1979 Yale University Press

Adrienne Vittadini © Graham Donald/CORBIS Sygma

Patricia Roberts Harris © Associated Press

1980: Sherry Lansing © Associated Press

Candy Lightner © Associated Press

Sister Elaine Roulet photograph by Harvey Wang, 2006

1981: Bette Bao Lord © Sophie Bassouls/CORBIS SYGMA

Jeane Kirkpatrick reprinted with the permission of The American Enterprise Institute for Public Policy Research, Washington, D.C.

Gerda Lerner photo by John Urban

Marcy Carsey © PACHA/CORBIS

Maya Lin © Associated Press

Vietnam Memorial Wall © Associated Press

Sandra Day O'Connor © Associated Press

Wilhelmina Holladay photograph by Sarah Hazlegrove, courtesy of the National Museum of Women in the Arts

1982: Buffy Sainte-Marie © Bettmann/CORBIS

Rebecca Matthias © Getty Images

Alice Walker © Associated Press

Ellen DeGeneres © Associated Press

Nancy Brinker © Associated Press

1983: Ellen Zwilich © Oscar White/CORBIS

Jenny Craig © PACHA/CORBIS

Katherine Dunham © Associated Press

Esther Dyson © Reuters/CORBIS

Lesley Stahl © Associated Press

Elizabeth Hanford Dole © Associated Press

Connie Chung © Getty Images

Sally Ride © Getty Images

Marion Ettlinger photograph by Marion Ettlinger

Barbara McClintock courtesy of the Library of Congress

1984: Sandra Cisneros © Getty Images

Catherine Crier © Getty Images

Donna Karan © Getty Images

Joan Benoit © Getty Images

Flossie Wong-Staal courtesy of Flossie Wong-Staal, Ph.D.

Diane Sawyer © Time & Life Pictures/Getty Images

Geraldine Ferraro © Getty Images

1985: E-volve! cover © 2004 Harvard Business School Press

Libby Riddles © Associated Press

Wilma Mankiller © Associated Press

Nancy Lopez © Associated Press

Vinita Gupta © AFP/Getty Images

Shannon Lucid © AFP/Getty Images

Oprah Winfrey photographs © Associated Press

1986: Gloria Estefan © AFP/Getty Images

Nien Cheng © Time & Life Pictures/Getty Images

American Girl doll © Time & Life Pictures/Getty Images

Rosie O'Donnell © Getty Images

Ann Bancroft © Reuters/CORBIS

Susan Butcher © Associated Press

Caryn Mandabach © Getty Images

Christa McAuliffe © Getty Images

Susan Polgar © Associated Press

1987: Camille Olivia Hanks Cosby © Getty Images

Gloria Anzaldua photo by Annie Valva. Nettie Lee Benson Latin American Collection, University of Texas Libraries, the University of Texas at Austin.

Katherine Siva Saubel courtesy of Katherine Siva Saubel

Jo Waldron © Associated Press

Anita Borg © Associated Press

Johnetta B. Cole © Getty Images

Mae Jemison © Time & Life Pictures/Getty Images

Nancy Pelosi © Getty Images

Sheri Poe © Time & Life Pictures/Getty Images

Donna Shalala © Getty Images

Aretha Franklin © Jerry Schatzberg/CORBIS

1988: Diane English © Getty Images

Marin Alsop © Associated Press

Gertrude Elion © Time & Life Pictures/Getty Images

Florence Griffith Joyner © AFP/Getty Images

Penny Marshall © Getty Images

Bonnie Blair © Getty Images

Ann Richards © Getty Images

Jackie Joyner-Kersee © AFP/Getty Images

Debi Thomas © AFP/Getty Images

1989: Sarah Brady © Getty Images

Amy Domini © Time & Life Pictures/Getty Images

Amy Tan © Getty Images

Julia Chang Bloch courtesy of Julia Chang Bloch

Judith Jamison © Associated Press

Steffie Allen courtesy of Steffie Allen

Wendy Wasserstein © Getty Images

1990: Sally Helgesen courtesy of Sally Helgesen

Whoopi Goldberg © AFP/Getty Images

Anne Wilson Schaef courtesy of Wilson Schaef Associates

Dr. Susan Love photograph courtesy of the Dr. Susan Love Research Foundation www.dslrf.org

Sylvia Alice Earle © Natalie B. Forbes/National Geographic Image Collection №638999

Antonia Novello © Getty Images

Ellen Ochoa © Getty Images

Katie Couric © Tim Graham/Getty Images

1991: Susan Faludi © Getty Images

Bernadine Healy © Time & Life Pictures/Getty Images

Audre Lorde © 2007 JEB (Joan E. Biren)

Cokie Roberts © Getty Images

Marilyn VanDerbur Atler © Bettmann/CORBIS

Sharon Pratt Dixon Kelly © Getty Images

Shulamit Ran © Associated Press

Jodie Foster © Getty Images

Anita Hill © Getty Images

Debora de Hoyos courtesy of Debora de Hoyos

1992: Gail Sheehy © Getty Images

Mary Fisher © Time & Life Pictures/Getty Images

Carol Bartz © Getty Images

1993: Rita Dove © Time & Life Pictures/Getty Images

Jocelyn Elders © AFP/Getty Images

Sheila W. Wellington courtesy of Sheila W. Wellington

Nora Ephron © Time & Life Pictures/Getty Images

Julie Krone © Getty Images

Ruth Bader Ginsburg © Time & Life Pictures/Getty Images

Janet Reno © AFP/Getty Images

Sheila Widnall © Associated Press

Glenna Goodacre courtesy of Glenna Goodacre

Condoleezza Rice © Getty Images

Toni Morrison © Time & Life Pictures/Getty Images

1994: Fran Lebowitz © Getty Images

Picabo Street © Getty Images

Myra Sadker courtesy of David Sadker

1995: Linda Chavez-Thompson © Getty Images

Eileen Collins © Getty Images

Ann Livermore © Getty Images

Ruth Simmons © Associated Press

Sheryl Swoopes © NBAE/Getty Images

Serena Williams © AFP/Getty Images

1996: Marion Downs courtesy of Marion Downs

Robin Roberts © Getty Images

1997: Swanee Hunt courtesy of Hunt Alternatives, LLC

Shelly Lazarus © Associated Press

Martha Stewart © Getty Images

Julie Taymor © Getty Images

Jody Williams © Getty Images

Violet Palmer © AFP/Getty Images

Madeleine Albright photograph © Getty Images

Jill Barad © Pascal Le Segretain/CORBIS SYGMA

1998: Geraldine Laybourne © Getty Images

Meg Whitman © Getty Images

Abby Joseph Cohen © Time & Life Pictures/Getty Images

Wilma Webb © Associated Press

Mia Hamm © Getty Images

1999: Cleo Parker Robinson © Associated Press

Carly Fiorina © Getty Images

Cathy L. Hughes © Getty Images

Andrea Jung © Time & Life Pictures/Getty Images

2000: The Women's Museum © Associated Press

Hillary Rodham Clinton © Getty Images

Susan Decker courtesy of Yahoo

2001: Gale Norton © Associated Press

Christine Todd Whitman © Getty Images

Colleen C. Barrett © NBAE/Getty Images

Regina Carter © Getty Images

2002: Anne M. Mulcahy © AFP/Getty Images

Patricia Russo © AFP/Getty Images

Halle Berry © AFP/Getty Images

Helen Greiner © Associated Press

Ann Moore © Getty Images

Karen Elliot House courtesy of *The Wall Street Journal*

2003: Mary Sammons © Time & Life Pictures/Getty Images

Sharon Allen © Najlah Feanny/Corbis

Ayanna Howard photo Courtesy of Human-Automation Systems Lab, Georgia Institute of Technology

Jane Friedman photo by George Lange, courtesy of HarperCollins *Publishers*

2004: Linda B. Buck © AFP/Getty Images

Susan Ivey courtesy of Susan Ivey

Sallie Krawcheck © Mark Peterson/CORBIS

Janet L. Robinson © Getty Images

Ann Sarnoff courtesy of Ann Sarnoff

Susan Lyne © Getty Images

Judy McGrath © Getty Images

Stephanie Burns © Associated Press

Susan Hockfield © Associated Press

Phylicia Rashad © Getty Images

Julie Theriot courtesy of Julie Theriot

2005: Brenda Barnes © Associated Press

Carolyn Vesper Bivens © Getty Images

Cristeta Comerford © Getty Images

2006: Mary Schapiro © Getty Images

Patricia Woertz © Kim Kulish/CORBIS

Paula Rosput Reynolds courtesy of Paula Rosput Reynolds

Indra K. Nooyi © Handout/epa/CORBIS

Katharine Jefferts Schori © AFP/Getty Images

2007: Drew Gilpin Faust by Kris Snibbe/Harvard News Office

Gail Kimbell © Associated Press

Page 230: Suffragists Celebrating 19th Amendment's Passage © Bettmann/CORBIS

Page 260: Belle Boyd courtesy of the Library of Congress

Amy Alcott © Tony Roberts/CORBIS

Minnie Maddern Fiske courtesy of the Library of Congress

Dian Fossey © Yann Arthus-Bertrand/CORBIS

Names Index

Abbott, Berenice, 1939
Abbott, Edith, 1908
Abbott, Grace, 1921
Abbott, Margaret, 1900
Abzug, Bella, 1971
Adams, Abigail, 1776
Adams, Maude, 1923
Addams, Jane, 1889
Adler, Stella, 1949
Agassiz, Elizabeth, 1882
Albright, Madeleine, 1997
Alcott, Amy, 1975
Alcott, Louisa May, 1868
Aldred, Anna Lee, 1939
Alexander, Hattie, 1937
Alexander, Sadie, 1935
Allen, Florence, 1922
Allen, Gracie, 1941
Allen, Sharon, 2003
Allen, Steffie, 1989
Alsop, Marin, 1988
Alvarado, Linda, 1976
Ancker-Johnson, Betsy, 1966
Anderson, Marian, 1939
Anderson, Mary, 1920
Anderson, Violette, 1926
Andrus, Ethel Percy, 1958
Angelou, Maya, 1972
Anthony, Susan B., 1869
Antin, Mary, 1912
Anzaldúa, Gloria, 1987
Apgar, Virginia, 1952

Arbus, Diane, 1972
Arbuthnot, May, 1940
Arden, Elizabeth, 1910
Arendt, Hannah, 1951
Ash, Mary Kay, 1963
Atler, Marilyn Van Derbur, 1991

Bacall, Lauren, 1944
Bachman, Maria Martin, 1833
Baez, Joan, 1960
Bagley, Sarah, 1844
Bailey, Pearl, 1970
Baker, Ella, 1957
Baker, Josephine, 1925
Balch, Emily Greene, 1946
Ball, Lucille, 1951
Bancroft, Ann, 1986
Banuelos, Romana, 1949
Barad, Jill, 1997
Barnes, Brenda, 2005
Barrett, Colleen C., 2001
Barrymore, Ethel, 1894
Barton, Clara, 1881
Bartz, Carol, 1992
Bascom, Florence, 1893
Bates, Katherine Lee, 1895
Bay, Josephine Perfect, 1956
Beach, Amy Marcy Cheney, 1896
Beard, Mary Ritter, 1946
Beaux, Cecilia, 1884
Beech, Olive Ann, 1950
Beecher, Catharine, 1852

Belmont, Alva Erskine Smith Vanderbilt, 1909
Bemis, Polly, 1872
Benedict, Ruth, 1934
Bennett, Sophia Hayden, 1886
Benoit, Joan, 1984
Berg, Gertrude, 1949
Berry, Halle, 2002
Bethune, Mary McLeod, 1904
Bickerdyke, Mary, 1862
Bigelow, Ruth, 1945
Birney, Alice McLellan, 1897
Bissell, Anna, 1889
Bissell, Emily, 1907
Bivens, Carolyn Vesper, 2005
Blackwell, Alice Stone, 1890
Blackwell, Antoinette Brown, 1853
Blackwell, Elizabeth, 1849
Blair, Bonnie, 1988
Blair, Emily Newell, 1921
Blanchard, Helen Augusta, 1873
Blatch, Harriet Stanton, 1902
Bloch, Julia Chang, 1989
Blodgett, Katherine, 1938
Bloomer, Amelia, 1851
Blum, Arlene, 1970
Blumkin, Rose, 1937
Bly, Nellie (Elizabeth Cochrane Seaman), 1889
Bombeck, Erma, 1968
Bonney, Thérèse, 1939
Bordas, Juana, 1977

Borg, Anita, 1987
Bourke-White, Margaret, 1936
Boyd, Belle, 1861
Boyle, Gertrude, 1972
Bradstreet, Anne, 1650
Bradwell, Myra, 1868
Brady, Sarah, 1989
Brathwaite-Burke, Yvonne, 1966
Breckenridge, Mary, 1925
Breckinridge, Sophonisba, 1901
Brent, Margaret, 1648
Brice, Fanny, 1921
Brinker, Nancy, 1982
Briones, Juana, 1844
Bronson, Ruth Muskrat, 1930
Brooks, Gwendolyn, 1949
Brown, Margaret Wise, 1937
Brown, Rachel Fuller, 1950
Brownmiller, Susan, 1975
Bryant, Lena, 1904
Buck, Linda B., 2004
Buck, Pearl S., 1938
Bullitt, Dorothy Stimson, 1947
Burnett, Carol, 1959
Burns, Lucy, 1912
Burns, Stephanie, 2004
Burroughs, Nannie Helen, 1900
Butcher, Susan, 1986
Butler, Octavia, 1979

Cabrini, Mother Frances Xavier, 1889
Calamity Jane (Martha Jane Cannary), 1870

Calderone, Mary Steichen, 1953
Caldicott, Helen, 1974
Caldwell, Sarah, 1976
Cannon, Annie Jump, 1896
Carnegie, Hattie, 1925
Carroll, Diahann, 1968
Carsey, Marcy, 1984
Carson, Rachel, 1962
Carter, Maybelle, 1939
Carter, Regina, 2001
Cary, Mary Ann Shadd, 1853
Cassatt, Mary, 1868
Cather, Willa, 1918
Catt, Carrie Chapman, 1900
Channing, Carol, 1964
Chapman, Maria Weston, 1832
Chappelle, Georgette (Dicky), 1943
Chavez-Thompson, Linda, 1995
Chen, Joyce, 1958
Cheng, Nien, 1986
Chesnut, Mary, 1905
Chiang, Anne, 1984
Chicago, Judy, 1979
Child, Julia, 1963
Child, Lydia Maria, 1833
Chisholm, Shirley, 1968
Chopin, Kate, 1889
Chung, Connie, 1983
Church, Ellen, 1930
Cisneros, Sandra, 1984
Claiborne Ortenberg, Elizabeth, 1976
Clark, Georgia Neese, 1949
Clark, Septima, 1920
Clarke, Edith, 1947
Claytor, Helen, 1967
Cleveland, Emeline Horton, 1855
Cline, Genevieve, 1928
Cline, Patsy, 1957
Clinton, Hillary Rodham, 2000
Coachman, Alice, 1948
Cobb, Jerrie, 1960
Coca, Imogene, 1925

Cochran, Jackie, 1942
Cochran, Josephine, 1886
Cohen, Abby Joseph, 1998
Colden, Jane, 1757
Cole, Johnetta B., 1987
Coleman, Bessie, 1921
Collins, Eileen, 1995
Colter, Mary, 1905
Comerford, Cristeta, 2005
Converse, Harriet Maxwell, 1891
Conway, Jill Ker, 1975
Cook, Betty, 1921
Cooney, Joan Ganz, 1969
Cooper, Anna Julia, 1892
Cope, Mother Marianne, 1888
Coppin, Fanny Jackson, 1869
Corbin, Margaret, 1779
Cori, Gerty Radnitz, 1947
Cornell, Katharine, 1930
Cosby, Camille Olivia Hanks, 1987
Coston, Martha, 1871
Couric, Katie, 1990
Cox, Marie, 1970
Craig, Jenny, 1983
Crandall, Prudence, 1833
Crawford, Joan, 1945
Crier, Catherine, 1984
Croly, Jane Cunningham, 1889
Crosby, Fanny, 1844
Crump, Diane, 1970
Crumpler, Rebecca Lee, 1864
Cruz, Celia, 1947
Cushman, Pauline, 1865

Dare, Virginia, 1587
Davis, Alice, 1922
Davis, Bette, 1929
Davis, Paulina Kellogg Wright, 1835
Dawidowicz, Lucy, 1975
Day, Dorothy, 1933
Decker, Susan, 2000
De Dominic, Patty, 1979

De Generes, Ellen, 1982
de Hoyos, Debora, 1991
De Kooning, Elaine, 1963
de Mille, Agnes, 1942
Demorest, Ellen Curtis, 1860
de Victor, Maude, 1978
de Wolfe, Elsie, 1913
Dick, Gladys, 1924
Dickinson, Emily, 1852
Didion, Joan, 1968
Dillman, Linda, 2003
Dix, Dorothea, 1843
Dodge, Grace Hoadley, 1886
Dodge, Mary Mapes, 1865
Dole, Elizabeth Hanford, 1983
Domini, Amy, 1989
Donahue, Margaret, 1950
Douglas, Jennie, 1862
Douglas, Marjory Stoneman, 1947
Dove, Rita, 1993
Downs, Marion, 1996
Drexel, Sister Mary Katharine, 1891
Duckworth, Ruth, 1964
Duncan, Isadora, 1917
Dunham, Katherine, 1983
Duniway, Abigail Scott, 1912
Dunn, Gertrude, 1952
Dworkin, Andrea, 1974
Dwyer, Doriot Anthony, 1952
Dyer, Mary, 1660
Dyson, Esther, 1983

Earhart, Amelia, 1932
Earle, Sylvia Alice, 1990
Eastman, Crystal, 1910
Eddy, Mary Baker, 1879
Edelman, Marian Wright, 1973
Ederle, Gertrude, 1926
Elder, Ruth, 1927
Elders, Jocelyn, 1993
Elion, Gertrude, 1988
Ellerbee, Linda, 1972

English, Diane, 1988
Ephron, Nora, 1993
Estefan, Gloria, 1986
Ettlinger, Marion, 1983
Eustis, Dorothy, 1929
Evans, Alice, 1917
Evert, Chris, 1974

Faludi, Susan, 1991
Farmer, Fannie, 1896
Farnham, Eliza Wood Burhans, 1864
Farrar, Margaret Petherbridge, 1924
Farrell, Suzanne, 1960
Fauset, Jessie Redmon, 1919
Faust, Drew Gilpin, 2007
Ferber, Edna, 1924
Ferguson, Catherine, 1793
Ferraro, Geraldine, 1984
Fertel, Ruth, 1965
Field, Pattie, 1925
Fields, Debbi, 1977
Fiorina, Carly, 1999
Fisher, Mary, 1992
Fiske, Minnie Maddern, 1882
Fitzgerald, Ella, 1934
Fleming, Peggy, 1968
Fleming, Williamina Stevens, 1906
Fletcher, Alice Cunningham, 1902
Flugge-Lotz, Irmgard, 1960
Flynn, Elizabeth Gurley, 1912
Follett, Mary Parker, 1896
Foltz, Clara Shortridge, 1878
Fonda, Jane, 1969
Ford, Betty, 1974
Fossey, Dian, 1967
Foster, Abby Kelley, 1838
Foster, Jodie, 1991
Frankenthaler, Helen, 1950
Franklin, Aretha, 1987
Freeman, Elizabeth ("Mum Bett"), 1781
Friedan, Betty, 1963
Friedman, Jane, 2003

Fudge, Ann, 1977
Fuller, Margaret, 1839
Fulton, Sarah, 1773
Furbish, Catherine, 1895

Gage, Matilda Joslyn, 1852
Garland, Judy, 1940
Genovese, Kitty, 1964
Gibbs, Katharine, 1911
Gibson, Althea, 1950
Gilbreth, Lillian, 1931
Gilman, Charlotte Perkins, 1892
Ginsberg, Sadie, 1957
Ginsburg, Ruth Bader, 1993
Glasgow, Ellen, 1925
Gleason, Kate, 1890
Goddard, Mary Katherine, 1775
Goeppert-Mayer, Maria, 1963
Goldberg, Whoopi, 1990
Goldman, Emma, 1889
Goodacre, Glenna, 1993
Goode, Sarah, 1885
Graham, Bette Nesmith, 1951
Graham, Katharine Meyer, 1974
Graham, Martha, 1927
Grasso, Ella, 1975
Gratz, Rebecca, 1838
Green, Edith, 1955
Green, Hetty, 1916
Greene, Catherine, 1807
Greiner, Helen, 2002
Grimké, Angelina, 1838
Grimké, Charlotte Forten, 1862
Grimké, Sarah, 1839
Gulick, Charlotte Vetter, 1910
Gupta, Vinita, 1985
Guthrie, Janet, 1977

Hale, Sarah Josepha, 1837
Hallaren, Mary Agnes, 1948
Hamer, Fannie Lou, 1964
Hamilton, Alice, 1947

Hamilton, Edith, 1930
Hamilton, Virginia, 1975
Hamm, Mia, 1999
Handler, Ruth, 1959
Hansberry, Lorraine, 1959
Harper, Frances E. W., 1860
Harper, Ida Husted, 1922
Harper, Martha Matilda, 1891
Harris, LaDonna, 1970
Harris, Patricia Roberts, 1979
Harrison, Hazel, 1915
Haughery, Margaret, 1840
Hawn, Goldie, 1968
Hayes, Helen, 1909
Haynes, Euphemia Lofton, 1943
Hazen, Elizabeth Lee, 1950
Healy, Bernadine, 1991
Helgesen, Sally, 1990
Hellman, Lillian, 1952
Henie, Sonja, 1927
Henry, Beulah, 1912
Hepburn, Katharine, 1933
Herrington, Alice, 1957
Hesse, Eva, 1972
Hicks, Beatrice, 1950
Higgins, Maggie, 1951
Hill, Anita, 1991
Hobby, Oveta Culp, 1942
Hockfield, Susan, 2004
Holdridge, Barbara, 1952
Holiday, Billie, 1935
Holladay, Wilhelmina, 1981
Holm, Jeanne M., 1971
Hopper, Grace Murray, 1943
Horne, Lena, 1943
Horne, Marilyn, 1969
Horney, Karen, 1939
Horwich, Frances, 1952
Hosmer, Harriet Goodhue, 1853
Houghton, Edith, 1946
House, Karen Elliot, 2002
Howard, Ayanna, 2003

Howe, Julia Ward, 1862
Howland, Emily, 1857
Hoxie, Vinnie Ream, 1866
Huerta, Dolores, 1962
Hughes, Cathy L., 1999
Hunt, Harriot Kezia, 1835
Hunt, Swanee, 1997
Hurston, Zora Neale, 1939
Husted, Marjorie Child, 1921
Hutchinson, Anne, 1637
Huxtable, Ada Louise, 1970

Idar, Jovita, 1911
Ivey, Susan, 2004

Jackson, Mahalia, 1934
Jackson, Shirley Ann, 1973
Jacob, Mary Phelps, 1914
Jacobi, Mary Putnam, 1872
Jacobs, Frances Wisebart, 1887
Jacobs, Jane, 1961
Jamison, Judith, 1989
Jemison, Mae, 1987
Jemison, Mary, 1758
Jenney, Mistress Sarah, 1644
Jewett, Sarah Orne, 1896
Johnson, Lady Bird, 1965
Johnson, Mary, 1622
Johnston, Frances Benjamin, 1889
Jones, Mother (Mary Harris), 1903
Jones, Sissieretta, 1888
Jordan, Barbara, 1966
Joyner, Florence Griffith, 1988
Joyner-Kersee, Jackie, 1988
Jung, Andrea, 1999

Kaahumanu, 1819
Kanter, Rosabeth Moss, 1985
Karan, Donna, 1984
Kassebaum, Nancy Landon, 1978
Keckley, Elizabeth, 1855
Keeler, Ruby, 1928

Keller, Helen, 1903
Kelley, Florence, 1899
Kelly, Sharon Pratt Dixon, 1991
Kelsey, Frances, 1962
Kennedy Onassis, Jacqueline, 1962
Key, Elizabeth, 1655
Kies, Mary, 1809
Kilgallen, Dorothy, 1936
Kimbell, Gail, 2007
King, Billie Jean, 1973
King, Carole, 1971
King, Coretta Scott, 1969
Kingston, Maxine Hong, 1976
Kirkpatrick, Jeane, 1981
Knight, Margaret, 1870
Knight, Sarah Kemble, 1704
Knox, Rose Markward, 1890
Koplovitz, Kay, 1977
Kovac, Carol, 2004
Krasner, Lee, 1951
Krawcheck, Sallie, 2004
Kreps, Juanita, 1972
Krone, Julie, 1993
Kübler-Ross, Elisabeth, 1969
Kuhn, Maggie, 1971
Kwolek, Stephanie, 1965

Ladd-Franklin, Christine, 1882
La Flesche Tibbles, Susette, 1880
Laney, Lucy Craft, 1890
Lange, Dorothea, 1936
Lansbury, Angela, 1944
Lansing, Sherry, 1980
Lathrop, Julia, 1912
Lauder, Estée, 1946
Laurie, Annie (Winifred Black), 1893
Laybourne, Geraldine, 1998
Lazarus, Emma, 1883
Lazarus, Shelly, 1997
Leavitt, Henrietta Swan, 1912
Lebowitz, Fran, 1994
Lee, Harper, 1960

Lee, Mother Ann, 1776
Lee, Rose Hum, 1956
Leibovitz, Annie, 1970
León, Tania, 1969
Lerner, Gerda, 1981
Lewis, Edmonia, 1867
Lewis, Ida, 1879
Lightner, Candy, 1980
Lin, Maya, 1984
Lindbergh, Anne Morrow, 1930
Little, Carole, 1975
Livermore, Ann, 1995
Livermore, Mary Rice, 1864
Locke, Patricia, 1978
Lockwood, Belva, 1879
Logan, Martha Daniell, 1754
Lombard, Carole, 1925
Loos, Anita, 1912
Lopez, Nancy, 1985
Lord, Bette Bao, 1984
Lorde, Audre (Gamba Adisa), 1991
Lotsey, Nancy, 1963
Love, Susan, 1990
Low, Juliette Gordon, 1912
Loy, Myrna, 1925
Luce, Clare Boothe, 1953
Lucid, Shannon, 1985
Ludington, Sybil, 1777
Luhan, Mabel Dodge, 1918
Lukens, Rebecca Webb, 1825
Lupino, Ida, 1933
Lyne, Susan, 2004
Lyon, Mary, 1837

MacKinnon, Catharine Alice, 1979
MacLaine, Shirley, 1955
Macy, Anne Sullivan, 1887
Madison, Dolley, 1814
Malone, Annie, 1902
Mandabach, Caryn, 1986
Mankiller, Wilma, 1985
Manley, Effa, 1935

Mansfield, Arabella, 1869
Mantell, Marianne, 1952
Marion, Frances, 1930
Marshall, Penny, 1988
Martin, Mary, 1949
Martinez, Maria Montoya, 1919
Masters, Sybilla, 1715
Matthias, Rebecca, 1982
Maxwell, Martha, 1876
McAuliffe, Christa, 1986
McBride, Mary Margaret, 1934
McClintock, Barbara, 1983
McClintock, Jessica, 1969
McDaniel, Hattie, 1940
McDowell, Anna Elizabeth, 1855
McGrath, Judy, 2004
McGrory, Mary, 1974
Mead, Margaret, 1928
Meadows, Jayne, 1946
Meir, Golda, 1969
Merman, Ethel, 1930
Meskis, Joyce, 1974
Messick, Dale, 1940
Midler, Bette, 1972
Millay, Edna St. Vincent, 1923
Millett, Kate, 1970
Miner, Myrtilla, 1851
Minoka-Hill, L. Rosa, 1899
Minor, Virginia, 1875
Missett, Judi Sheppard, 1969
Mitchell, Margaret, 1936
Mitchell, Maria, 1848
Monroe, Marilyn, 1946
Montgomery, Mary Jane, 1864
Moody, Helen Wills, 1924
Moody, Lady Deborah, 1645
Moore, Ann, 2002
Moore, Marianne, 1915
Moore, Mary Tyler, 1970
Moorehead, Agnes, 1929
Moreno, Rita, 1962
Morgan, Julia, 1898

Morrison, Toni, 1993
Moses, Grandma, 1939
Moskowitz, Belle, 1913
Motley, Constance Baker, 1943
Mott, Lucretia, 1840
Mourning Dove, 1927
Mulcahy, Anne M., 2002
Muldowney, Shirley, 1965
Murray, Judith Sargent Stevens, 1790
Murray, Pauli, 1944
Musgrove, Mary, 1733

Nampeyo, 1895
Nation, Carry, 1899
Natori, Josie Cruz, 1977
Navratilova, Martina, 1975
Neakok, Sadie, 1960
Nealey, Bertha, 1945
Neal, Patricia, 1964
Nelson, Ida Gray, 1890
Nestor, Agnes, 1902
Nevelson, Louise, 1940
Nichols, Ruth, 1932
Nidetch, Jean, 1963
Ninham, Sandra, 1976
Noether, Amalie Emmy, 1921
Nooyi, Indra K., 2006
Norton, Eleanor Holmes, 1970
Norton, Gale, 2001
Notestein, Ada Comstock, 1923
Novello, Antonia, 1990

Oakley, Annie (Phoebe Ann Mosey), 1885
Oates, Joyce Carol, 1970
Ochoa, Ellen, 1990
O'Connor, Flannery, 1952
O'Connor, Sandra Day, 1981
O'Donnell, Rosie, 1986
O'Keeffe, Georgia, 1916
Olsen, Tillie, 1961
Pierce, Charlotte Woodward, 1920
O'Neil, Kitty, 1976
Ono, Yoko, 1964

O'Sullivan, Mary Kenney, 1892
Outerbridge, Mary Ewing, 1874
Ovington, Mary White, 1909

Palmer, Bertha, 1892
Palmer, Frances (Fanny) Bond, 1849
Palmer, Phoebe Worrall, 1837
Palmer, Violet, 1997
Park, Maud Wood, 1901
Parker, Dorothy, 1927
Parks, Rosa, 1955
Parrish, Essie, 1941
Parsons, Elsie Clews, 1939
Parsons, Lucy (Lucia González), 1886
Parton, Dolly, 1975
Patrick, Ruth, 1933
Patterson, Mary Jane, 1862
Paul, Alice, 1912
Pauley, Jane, 1976
Peabody, Elizabeth, 1860
Peale, Anna Claypoole, 1824
Peale, Sarah Miriam, 1824
Pearce, Louise, 1919
Pearl, Minnie, 1940
Peck, Annie Smith, 1911
Pelosi, Nancy, 1987
Peña, Soledad, 1911
Penn, Hannah Callowhill, 1712
Pennington, Mary Engle, 1895
Perkins, Frances, 1933
Perry, Lilla Cabot, 1889
Peters, Roberta, 1950
Peterson, Esther, 1961
Philipse, Margaret Hardenbroeck, 1660
Phillips, Julia, 1973
Pickersgill, Mary Young, 1813
Pickford, Mary, 1916
Picon, Molly, 1903
Picotte, Susan La Flesche, 1889
Pierce, Charlotte Woodward, 1920
Pinckney, Eliza Lucas, 1744
Pinkham, Lydia Estes, 1875

Pitcher, Molly (Mary McCauley), 1778
Pocahontas, 1608
Poe, Sheri, 1987
Pogrebin, Letty Cottin, 1971
Polgar, Susan, 1986
Pool, Judith Graham, 1965
Porter, Katherine Anne, 1962
Porter, Sylvia, 1942
Post, Emily, 1922
Post, Marjorie Merriweather, 1929
Powell, Maud, 1885
Price, Florence Beatrice Smith, 1933
Price, Leontyne, 1952
Priesand, Sally J., 1972
Prince, Lucy Terry, 1746

Qoyawayma, Polingaysi, 1925
Queler, Eve, 1966
Quimby, Harriet, 1911

Radner, Gilda, 1975
Rainey, Ma, 1923
Ran, Shulamit, 1991
Rand, Ayn, 1943
Rankin, Jeannette, 1917
Rashad, Phylicia, 2004
Rawlings, Marjorie Kinnan, 1938
Ray, Charlotte, 1872
Ray, Dixy Lee, 1976
Reed, Esther DeBerdt, 1780
Reno, Janet, 1993
Resnick, Judith, 1978
Reynolds, Malvina, 1950
Reynolds, Paula Rosput, 2006
Rice, Condoleezza, 1993
Rich, Adrienne, 1973
Richards, Ann, 1988
Richards, Ellen Swallow, 1882
Riddles, Libby, 1985
Ride, Sally, 1983
Rivera, Chita, 1957
Rivers, Joan, 1975

Roberts, Cokie, 1991
Roberts, Robin, 1996
Robinson, Cleo Parker, 1999
Robinson, Janet L., 2004
Robinson, Julia, 1976
Roche, Josephine, 1912
Rockwell, Mabel MacFerran, 1935
Rogers, Edith Nourse, 1925
Rogers, Ginger, 1935
Rolf, Ida, 1967
Roosevelt, Eleanor, 1933
Rose, Ernestine, 1840
Rosenthal, Ida, 1923
Ross, Betsy, 1776
Roulet, Sister Elaine, 1980
Rowland, Pleasant T., 1986
Rowlandson, Mary, 1682
Rowson, Susanna Haswell, 1790
Rubin, Vera, 1970
Rubinstein, Helena, 1914
Rudkin, Margaret, 1936
Rudolph, Wilma, 1960
Ruffin, Josephine St. Pierre, 1894
Rukeyser, Muriel, 1935
Russell, Rosalind, 1934
Russo, Patricia, 2002

Sabin, Florence, 1925
Sacagawea, 1805
Sadker, Myra, 1994
St. Denis, Ruth, 1915
Sainte-Marie, Buffy, 1982
Sammons, Mary, 2003
Sampson, Deborah, 1782
Sandoz, Mari, 1935
Sanger, Margaret, 1916
Sarnoff, Ann, 2004
Satir, Virginia, 1964
Saubel, Katherine Siva, 1987
Sawyer, Diane, 1984
Sawyer, Ruth, 1915
Schaef, Anne Wilson, 1990

Schapiro, Mary, 2006
Schneiderman, Rose, 1903
Schori, Katharine Jefferts, 2006
Schroeder, Becky, 1974
Schroeder, Patricia, 1972
Schulze, Tye Leung, 1910
Schwartz, Felice, 1962
Seeger, Ruth, 1930
Seibert, Florence, 1941
Semple, Ellen Churchill, 1921
Serrano, Lupe, 1953
Seton, Elizabeth Ann Bayley, 1797
Shalala, Donna, 1987
Shambaugh, Jessie Field, 1901
Shaw, Anna Howard, 1880
Sheehy, Gail, 1992
Shore, Dinah, 1955
Shriver, Eunice Kennedy, 1961
Siebert, Muriel, 1967
Silko, Leslie, 1977
Silkwood, Karen, 1974
Sills, Beverly, 1955
Simmons, Amelia, 1796
Simmons, Ruth, 1995
Sinclair, Mary, 1967
Slye, Maud, 1911
Smith, Bessie, 1919
Smith, Kate, 1926
Smith, Margaret Bayard, 1824
Smith, Margaret Chase, 1940
Smith, Sophia, 1871
Sobrino, Maria de Lourdes, 1982
Solomon, Hannah G., 1893
Sontag, Susan, 1966
Stahl, Leslie, 1983
Stamberg, Susan, 1972
Stanton, Elizabeth Cady, 1848
Stanwyck, Barbara, 1973
Stein, Gertrude, 1909
Steinem, Gloria, 1971
Stephens, Helen, 1936
Stern, Catherine, 1938

Stevens, Nettie, 1905
Stewart, Maria, 1831
Stewart, Martha, 1997
Stone, Lucy, 1850
Stowe, Harriet Beecher, 1852
Streep, Meryl, 1977
Street, Picabo, 1994
Streisand, Barbra, 1962
Strong, Harriet, 1887
Swanson, Gloria, 1927
Switzer, K. (Kathrine), 1967
Swoopes, Sheryl, 1995
Szold, Henrietta, 1912

Tallchief, Maria, 1954
Tan, Amy, 1989
Tandy, Jessica, 1947
Tarbell, Ida, 1904
Taussig, Helen Brooke, 1944
Taylor, Elizabeth, 1942
Taylor, Lucy, 1866
Taymor, Julie, 1997
Tekawitha, Kateri, 1676
Temple, Shirley, 1935
Tenayuca, Emma, 1938
Tenderich, Gertrude, 1934
Terrell, Mary Church, 1896
Terry, Hilda, 1941
Tharp, Twyla, 1965
Thering, Sister Rose, 1957
Theriot, Julie, 2004
Thomas, Alma, 1924
Thomas, Debi, 1988
Thomas, Helen, 1974
Thomas, Marlo, 1966
Thomas, M. Carey, 1924
Thompson, Dorothy, 1936
Tibbles, Susette La Flesche, 1880
Tomlin, Lily, 1977
Tompkins, Sally, 1861
Towner, Margaret, 1956
Truitt, Anne, 1973

Truth, Sojourner, 1843
Tubman, Harriet, 1850
Tuchman, Barbara, 1962
Tucker, Sophie, 1906
Turner, Tina, 1956

Uchida, Yoshiko, 1949
Urso, Camilla, 1852

Vaughan, Sarah, 1942
Vaught, Wilma, 1966
Vernon, Lillian, 1954
Vittadini, Adrienne, 1979

Wald, Lillian D., 1895
Waldron, Jo, 1987
Walker, Alice, 1982
Walker, Madam C. J., 1905
Walker, Maggie Lena, 1903
Walker, Mary Edwards, 1860
Wallace, Lila, 1922
Walters, Barbara, 1976
Ward, Nancy, 1755
Warren, Mercy Otis, 1773
Wasserstein, Wendy, 1989
Waters, Ethel, 1927
Wattleton, Faye, 1978
Wauneka, Annie Dodge, 1963
Webb, Wilma, 1998
Webster, Alma, 1976
Weddington, Sarah, 1973
Wellington, Sheila, 1993
Wells-Barnett, Ida B., 1895
Welty, Eudora, 1972
West, Mae, 1914
Westheimer, Ruth, 1970
Wharton, Edith, 1920
Wheatley, Phillis, 1773
White, Betty, 1953
White, Ellen, 1860
Whitehead, Nancy Dickerson, 1960
Whitener, Catherine Evans, 1895

Whitman, Christine Todd, 2004
Whitman, Meg, 1998
Whitman, Narcissa Prentiss, 1836
Widnall, Sheila, 1993
Wilder, Laura Ingalls, 1935
Willard, Emma, 1824
Willard, Frances, 1874
Williams, Jody, 1997
Williams, Serena, 1995
Wilson, Harriet, 1859
Winfrey, Oprah, 1985
Winnemucca, Sarah, 1883
Winters, Shelley, 1954
Wise, Brownie, 1954
Woertz, Patricia, 2006
Wolcott, Marion Post, 1938
Wong, Jade Snow, 1950
Wong-Staal, Flossie, 1984
Woodhull, Victoria, 1872
Woolley, Mary Emma, 1894
Workman, Fanny Bullock, 1906

Wright, Frances (Fanny), 1829
Wright, Martha Coffin, 1833
Wright, Patience, 1769
Wu, Chien Shiung, 1944
Wylie, Mary Washington, 1968

Yalow, Rosalyn, 1977
Yamamoto, Hisaye, 1950
Yu, Alice Fong, 1926

Zaharias, Babe Didrikson, 1950
Zakrzewska, Marie Elizabeth, 1862
Zitkala Sa (Red Bird; Gertrude Bonnin),
    1904
Zwilich, Ellen, 1983

Professions Index

ABOLITIONISTS

Cary, Mary Ann Shadd, 1853
Chapman, Maria Weston, 1832
Child, Lydia Maria, 1833
Davis, Paulina Kellogg Wright, 1835
Foster, Abby Kelley, 1838
Grimké, Angelina, 1838
Grimké, Sarah, 1839
Harper, Frances E. W., 1860
Howland, Emily, 1857
Miner, Myrtilla, 1854
Mott, Lucretia, 1840
Stowe, Harriet Beecher, 1852
Truth, Sojourner, 1843
Tubman, Harriet, 1850
Wright, Martha Coffin, 1833

ACADEMICS

Agassiz, Elizabeth, 1882
Arendt, Hannah, 1954
Cole, Johnetta B., 1987
Conway, Jill Ker, 1975
Cooper, Anna Julia, 1892
Faust, Drew Gilpin, 2007
Hockfield, Susan, 2004
Hopper, Grace Murray, 1943
Hunt, Swanee, 1997
Jackson, Shirley Ann, 1973
Jacobs, Jane, 1961
Kanter, Rosabeth Moss, 1985
Lee, Rose Hum, 1956

Lerner, Gerda, 1984
Locke, Patricia, 1978
Mansfield, Arabella, 1869
Mitchell, Maria, 1848
Notestein, Ada Comstock, 1923
Rice, Condoleezza, 1993
Sabin, Florence, 1925
Shalala, Donna, 1987
Simmons, Ruth, 1995
Thomas, M. Carey, 1924
Woolley, Mary Emma, 1891
See also Educators

ACTIVISTS

Andrus, Ethel Percy, 1958
Baez, Joan, 1960
Baker, Ella, 1957
Balch, Emily Greene, 1946
Bethune, Mary McLeod, 1904
Blackwell, Alice Stone, 1890
Blair, Emily Newell, 1924
Blatch, Harriet Stanton, 1902
Bordas, Juana, 1977
Brady, Sarah, 1989
Burns, Lucy, 1912
Burroughs, Nannie Helen, 1900
Caldicott, Helen, 1971
Clark, Septima, 1920
Claytor, Helen, 1967
Cole, Johnetta B., 1987
Converse, Harriet Maxwell, 1891

Cox, Marie, 1970
Davis, Alice, 1922
de Victor, Maude, 1978
Downs, Marion, 1996
Duniway, Abigail Scott, 1912
Edelman, Marian Wright, 1973
Fisher, Mary, 1992
Fonda, Jane, 1969
Friedan, Betty, 1963
Goldman, Emma, 1889
Hamer, Fannie Lou, 1964
Harris, LaDonna, 1970
Herrington, Alice, 1957
Hill, Anita, 1991
Huerta, Dolores, 1962
Kelley, Florence, 1899
Kuhn, Maggie, 1971
Lathrop, Julia, 1912
Lightner, Candy, 1980
Locke, Patricia, 1978
Lorde, Audre (Gamba Adisa), 1991
Love, Susan, 1990
Mankiller, Wilma, 1985
Millett, Kate, 1970
"Mum Bett" (Elizabeth Freeman), 1781
Murray, Pauli, 1944
Nation, Carry, 1899
Ovington, Mary White, 1909
Park, Maud Wood, 1901
Parks, Rosa, 1955
Parsons, Lucy (Lucia González), 1886

Paul, Alice, 1912
Peterson, Esther, 1961
Pogebrin, Letty Cottin, 1971
Rankin, Jeannette, 1917
Roulet, Sister Elaine, 1980
Sadker, Myra, 1994
Sanger, Margaret, 1916
Schwartz, Felice, 1962
Silkwood, Karen, 1974
Sinclair, Mary, 1967
Steinem, Gloria, 1971
Tenayuca, Emma, 1938
Terrell, Mary Church, 1896
Tibbles, Susette La Flesche, 1880
Waldron, Jo, 1987
Walker, Mary Edwards, 1860
Wauneka, Annie Dodge, 1963
Wells-Barnett, Ida B., 1895
White, Betty, 1953
Willard, Frances, 1874
Williams, Jody, 1997
Winnemucca, Sarah, 1883
See also specific causes

ACTORS

Adams, Maude, 1923
Bacall, Lauren, 1944
Ball, Lucille, 1951
Barrymore, Ethel, 1894
Berry, Halle, 2002
Brice, Fanny, 1921

Carroll, Diahann, 1968
Cornell, Katharine, 1930
Crawford, Joan, 1945
Davis, Bette, 1929
Fiske, Minnie Maddern, 1882
Fonda, Jane, 1969
Foster, Jodie, 1991
Garland, Judy, 1940
Goldberg, Whoopi, 1990
Hawn, Goldie, 1968
Hayes, Helen, 1909
Hepburn, Katharine, 1933
Keeler, Ruby, 1928
Lansbury, Angela, 1944
Lombard, Carole, 1925
Loy, Myrna, 1925
Lupino, Ida, 1933
MacLaine, Shirley, 1955
Martin, Mary, 1949
McDaniel, Hattie, 1940
Meadows, Jayne, 1946
Merman, Ethel, 1930
Monroe, Marilyn, 1946
Moore, Mary Tyler, 1970
Moorehead, Agnes, 1929
Neal, Patricia, 1964
Pickford, Mary, 1916
Picon, Molly, 1903
Rashad, Phylicia, 2004
Rivera, Chita, 1957
Rogers, Ginger, 1935
Rowson, Susanna Haswell, 1790
Russell, Rosalind, 1934
Stanwyck, Barbara, 1973
Streep, Meryl, 1977
Streisand, Barbra, 1962
Swanson, Gloria, 1927
Tandy, Jessica, 1947
Taylor, Elizabeth, 1942
Temple Black, Shirley, 1935
Thomas, Marlo, 1966
Tomlin, Lily, 1977
Tucker, Sophie, 1906

West, Mae, 1944
White, Betty, 1953
Winfrey, Oprah, 1985
Winters, Shelley, 1954

ADVENTURERS
Bancroft, Ann, 1986
Blum, Arlene, 1970
Calamity Jane (Martha Jane Cannary), 1870
Oakley, Annie (Phoebe Ann Mosey), 1885
Peck, Annie Smith, 1911
Sacagawea, 1805
Whitman, Narcissa Prentiss, 1836
Workman, Fanny Bullock, 1906

ANTHROPOLOGISTS
Benedict, Ruth, 1934
Dunham, Katherine, 1983
Fletcher, Alice Cunningham, 1902
Mead, Margaret, 1928
Parrish, Essie, 1944
Parsons, Elsie Clews, 1939
Saubel, Katherine Siva, 1987

ARCHITECTS
Bennett, Sophia Hayden, 1886
Colter, Mary, 1905
Goodacre, Glenna, 1993
Lin, Maya, 1984
Morgan, Julia, 1898

ARTISTS.
See Visual Artists

ASTRONAUTS
Collins, Eileen, 1995
Jemison, Mae, 1987
Lucid, Shannon, 1985
McAuliffe, Christa, 1986
Ochoa, Ellen, 1990
Resnick, Judith, 1978
Ride, Sally, 1983

ASTRONOMERS
Cannon, Annie Jump, 1896
Fleming, Williamina Stevens, 1906
Leavitt, Henrietta Swan, 1912
Mitchell, Maria, 1848
Rubin, Vera, 1970

ATHLETES
Abbott, Margaret, 1900
Alcott, Amy, 1975
Aldred, Anna Lee, 1939
Benoit, Joan, 1984
Blair, Bonnie, 1988
Butcher, Susan, 1986
Coachman, Alice, 1948
Crump, Diane, 1970
Dunn, Gertrude, 1952
Ederle, Gertrude, 1926
Evert, Chris, 1974
Fleming, Peggy, 1968
Gibson, Althea, 1950
Guthrie, Janet, 1977
Hamm, Mia, 1999
Henie, Sonja, 1927
Houghton, Edith, 1946
Joyner, Florence Griffith, 1988
Joyner-Kersee, Jackie, 1988
King, Billie Jean, 1973
Krone, Julie, 1993
Lopez, Nancy, 1985
Lotsey, Nancy, 1963
Moody, Helen Wills, 1924
Muldowney, Shirley, 1965
Navratilova, Martina, 1975
O'Neil, Kitty, 1976
Outerbridge, Mary Ewing, 1874
Palmer, Violet, 1997
Riddles, Libby, 1985
Roberts, Robin, 1996
Rudolph, Wilma, 1960
Stephens, Helen, 1936
Street, Picabo, 1994
Switzer, K. (Kathrine), 1967

Swoopes, Sheryl, 1995
Thomas, Debi, 1988
Williams, Serena, 1995
Zaharias, Babe Didrikson, 1950

ATTORNEYS. See Lawyers

AVIATORS
Cobb, Jerrie, 1960
Cochran, Jackie, 1942
Coleman, Bessie, 1924
Collins, Eileen, 1995
Earhart, Amelia, 1932
Elder, Ruth,, 1927
Lindbergh, Anne Morrow, 1930
Nichols, Ruth, 1932
Quimby, Harriet, 1911

BIOLOGISTS
Carson, Rachel, 1962
Dick, Gladys, 1924
Earle, Sylvia Alice, 1990
Evans, Alice, 1917
Stevens, Nettie, 1905
Wong-Staal, Flossie, 1984

BOTANISTS AND
HORTICULTURISTS
Colden, Jane, 1757
Furbish, Catherine, 1895
Logan, Martha Daniell, 1754

BUSINESS PROFESSIONALS. See
Corporate Executives; Entrepreneurs

CABINET MEMBERS
Albright, Madeleine, 1997
Dole, Elizabeth Hanford, 1983
Harris, Patricia Roberts, 1979
Kreps, Juanita, 1972
Norton, Gale, 2001
Perkins, Frances, 1933
Reno, Janet, 1993
Rice, Condoleezza, 1993

Shalala, Donna, 1987
Widnall, Sheila, 1993

CARTOONISTS
Messick, Dale, 1940
Terry, Hilda, 1941

CHEFS
Chen, Joyce, 1958
Child, Julia, 1963
Comerford, Cristeta, 2005
See also Cookbook Authors

CHEMISTS
Burns, Stephanie, 2004
Cori, Gerty Radnitz, 1947
Kwolek, Stephanie, 1965
Pennington, Mary Engle, 1895
Richards, Ellen Swallow, 1882
Theriot, Julie, 2004

CHOREOGRAPHERS. *See Dancers and Choreographers*

CIVIC LEADERS
Barton, Clara, 1884
Birney, Alice McLellan, 1897
Claytor, Helen, 1967
Cook, Betty, 1924
Ovington, Mary White, 1909
Palmer, Bertha, 1892
Ruffin, Josephine St. Pierre, 1894
Sanger, Margaret, 1916
Shambaugh, Jessie Field, 1901
Szold, Henrietta, 1912
See also Public Servants; Social Reformers

CIVIL RIGHTS ADVOCATES
Alexander, Sadie, 1935
Baker, Ella, 1957
Clark, Septima, 1920
Coleman, Bessie, 1924
Fauset, Jessie Redmon, 1919

Hamer, Fannie Lou, 1964
Harris, LaDonna, 1970
King, Coretta Scott, 1969
"Mum Bett" (Elizabeth Freeman), 1781
Ovington, Mary White, 1909
Parks, Rosa, 1955
Terrell, Mary Church, 1896
Wells-Barnett, Ida B., 1895

CIVIL WAR PERSONALITIES
Bickerdyke, Mary, 1862
Boyd, Belle, 1861
Chesnut, Mary, 1905
Coston, Martha, 1871
Cushman, Pauline, 1865
Grimké, Charlotte Forten, 1862
Howe, Julia Ward, 1862
Livermore, Mary Rice, 1864
Tompkins, Sally, 1861
Walker, Mary Edwards, 1860

COLONIAL ERA PERSONALITIES
Bradstreet, Anne, 1650
Brent, Margaret, 1648
Colden, Jane, 1757
Dare, Virginia, 1587
Dyer, Mary, 1660
Goddard, Mary Katherine, 1775
Hutchinson, Anne, 1637
Jemison, Mary, 1758
Jenney, Mistress Sarah, 1644
Johnson, Mary, 1622
Key, Elizabeth, 1655
Knight, Sarah Kemble, 1704
Logan, Martha Daniell, 1751
Masters, Sybilla, 1715
Moody, Lady Deborah, 1645
Musgrove, Mary, 1733
Penn, Hannah Callowhill, 1712
Philipse, Margaret Hardenbroeck, 1660
Pinckney, Eliza Lucas, 1744

Pocahontas, 1608
Prince, Lucy Terry, 1746
Ross, Betsy, 1776
Rowlandson, Mary, 1682
Tekawitha, Kateri, 1676
Ward, Nancy, 1755
Warren, Mercy Otis, 1773
Wheatley, Phillis, 1773
Wright, Patience, 1769

COMPOSERS
Beach, Amy Marcy Cheney, 1896
León, Tania, 1969
Ono, Yoko, 1964
Price, Florence Beatrice Smith, 1933
Ran, Shulamit, 1991
Seeger, Ruth, 1930
Zwilich, Ellen, 1983

COMPUTER SCIENCE AND INDUSTRY LEADERS
Bartz, Carol, 1992
Borg, Anita, 1987
Dyson, Esther, 1983
Fiorina, Carly, 1999
Gupta, Vinita, 1985
Hopper, Grace Murray, 1943
Kovac, Carol, 2004
Livermore, Ann, 1995

CONDUCTORS
Alsop, Marin, 1988
Caldwell, Sarah, 1976
León, Tania, 1969
Queler, Eve, 1966

CONGRESSIONAL MEMBERS. *See Legislators; Senators*

COOKBOOK AUTHORS
Chen, Joyce, 1958
Child, Julia, 1963
Farmer, Fannie, 1896

Husted, Marjorie Child, 1924
Simmons, Amelia, 1796

CORPORATE EXECUTIVES
Allen, Sharon, 2003
Ash, Mary Kay, 1963
Barad, Jill, 1997
Barnes, Brenda, 2005
Barrett, Colleen C., 2001
Bartz, Carol, 1992
Bay, Josephine Perfect, 1956
Bissell, Anna, 1889
Blumkin, Rose, 1937
Boyle, Gertrude, 1972
Bullitt, Dorothy Stimson, 1947
Burns, Stephanie, 2004
Carsey, Marcy, 1981
Cohen, Abby Joseph, 1998
Decker, Susan, 2000
Dillman, Linda, 2003
Dyson, Esther, 1983
Fertel, Ruth, 1965
Fiorina, Carly, 1999
Friedman, Jane, 2003
Fudge, Ann, 1977
Gleason, Kate, 1890
Handler, Ruth, 1959
House, Karen Elliot, 2002
Ivey, Susan, 2004
Jung, Andrea, 1999
Knox, Rose Markward, 1890
Koplovitz, Kay, 1977
Kovac, Carol, 2004
Krawcheck, Sallie, 2004
Lansing, Sherry, 1980
Lauder, Estée, 1946
Laybourne, Geraldine, 1998
Lazarus, Shelly, 1997
Livermore, Ann, 1995
Lyne, Susan, 2004
Manley, Effa, 1935
McGrath, Judy, 2004
Moore, Ann, 2002
Mulcahy, Anne M., 2002

Nooyi, Indra K., 2006
Reynolds, Paula Rosput, 2006
Robinson, Janet L., 2004
Rosenthal, Ida, 1923
Rudkin, Margaret, 1936
Russo, Patricia, 2002
Sammons, Mary, 2003
Sarnoff, Ann, 2004
Schapiro, Mary, 2006
Siebert, Muriel, 1967
Tenderich, Gertrude, 1934
Vernon, Lillian, 1951
Walker, Madam C. J., 1905
Whitman, Meg, 1998
Winfrey, Oprah, 1985
Woertz, Patricia, 2006
See also Entrepreneurs; Sports Executives

DANCERS AND CHOREOGRAPHERS
Baker, Josephine, 1925
de Mille, Agnes, 1942
Duncan, Isadora, 1917
Dunham, Katherine, 1983
Farrell, Suzanne, 1960
Graham, Martha, 1927
Jamison, Judith, 1989
Keeler, Ruby, 1928
Rivera, Chita, 1957
Robinson, Cleo Parker, 1999
Rogers, Ginger, 1935
St. Denis, Ruth, 1915
Serrano, Lupe, 1953
Tallchief, Maria, 1954
Tharp, Twyla, 1965

DENTISTS
Nelson, Ida Gray, 1890
Taylor, Lucy, 1866

DIPLOMATS
Albright, Madeleine, 1997
Bloch, Julia Chang, 1989
Field, Pattie, 1925

Hunt, Swanee, 1997
Kirkpatrick, Jeane, 1981
Luce, Clare Boothe, 1953
Temple Black, Shirley, 1935
Woolley, Mary Emma, 1891

DOCTORS. *See Physicians*

EDITORS
Cary, Mary Ann Shadd, 1853
Child, Lydia Maria, 1833
Fauset, Jessie Redmon, 1919
Fuller, Margaret, 1839
Graham, Katharine Meyer, 1974
Hale, Sarah Josepha, 1837
McDowell, Anna Elizabeth, 1855
Moore, Marianne, 1915
Pogrebin, Letty Cottin, 1971
Steinem, Gloria, 1971

EDUCATORS
Abbott, Berenice, 1939
Adler, Stella, 1949
Agassiz, Elizabeth, 1882
Andrus, Ethel Percy, 1958
Beecher, Catharine, 1852
Bethune, Mary McLeod, 1904
Cary, Mary Ann Shadd, 1853
Cole, Johnetta B., 1987
Comstock Notestein, Ada, 1923
Conway, Jill Ker, 1975
Cooper, Anna Julia, 1892
Coppin, Fanny Jackson, 1869
Crandall, Prudence, 1833
Faust, Drew Gilpin, 2007
Ferguson, Catherine, 1793
Flugge-Lotz, Irmgard, 1960
Gibbs, Katharine, 1911
Ginsberg, Sadie, 1957
Gratz, Rebecca, 1838
Grimké, Charlotte Forten, 1862
Hockfield, Susan, 2004
Horwich, Frances, 1952

Howland, Emily, 1857
Hunt, Swanee, 1997
Jackson, Shirley Ann, 1973
Kanter, Rosabeth Moss, 1985
Laney, Lucy Craft, 1890
Lee, Rose Hum, 1956
Locke, Patricia, 1978
Lyon, Mary, 1837
Macy, Anne Sullivan, 1887
Malone, Annie, 1902
Mansfield, Arabella, 1869
McAuliffe, Christa, 1986
Miner, Myrtilla, 1851
Mitchell, Maria, 1848
Neakok, Sadie, 1960
Patterson, Mary Jane, 1862
Peabody, Elizabeth, 1860
Qoyawayma, Polingaysi, 1925
Rice, Condoleezza, 1993
Richards, Ellen Swallow, 1882
Sabin, Florence, 1925
Seton, Elizabeth Ann Bayley, 1797
Shalala, Donna, 1987
Simmons, Ruth, 1995
Stern, Catherine, 1938
Thomas, Alma, 1924
Thomas, M. Carey, 1924
Willard, Emma, 1821
Woolley, Mary Emma, 1891
Yu, Alice Fong, 1926

ENGINEERS
Clarke, Edith, 1947
Gilbreth, Lillian, 1931
Gleason, Kate, 1890
Greiner, Helen, 2002
Hicks, Beatrice, 1950
Howard, Ayanna, 2003
Morgan, Julia, 1898
Ochoa, Ellen, 1990
Resnick, Judith, 1978
Rockwell, Mabel MacFerran, 1935
Widnall, Sheila, 1993

ENTERTAINERS
Allen, Gracie, 1944
Bailey, Pearl, 1970
Baker, Josephine, 1925
Berg, Gertrude, 1949
Brice, Fanny, 1921
Burnett, Carol, 1959
Channing, Carol, 1964
Coca, Imogene, 1925
De Generes, Ellen, 1982
Goldberg, Whoopi, 1990
Lebowitz, Fran, 1994
Martin, Mary, 1949
Midler, Bette, 1972
Moreno, Rita, 1962
Oakley, Annie (Phoebe Ann Mosey), 1885
O'Donnell, Rosie, 1986
Parton, Dolly, 1975
Pearl, Minnie, 1940
Radner, Gilda, 1975
Rivers, Joan, 1975
Tomlin, Lily, 1977
See also Actors; Musicians; Television and Radio Personalities; Singers

ENTREPRENEURS
Allen, Steffie, 1989
Alvarado, Linda, 1976
Arden, Elizabeth, 1910
Ash, Mary Kay, 1963
Ball, Lucille, 1951
Banuelos, Romana, 1949
Beech, Olive Ann, 1950
Bemis, Polly, 1872
Bigelow, Ruth, 1945
Bissell, Anna, 1889
Blumkin, Rose, 1937
Briones, Juana, 1844
Bryant, Lena, 1904
Bullitt, Dorothy Stimson, 1947
Carsey, Marcy, 1984
Claiborne Ortenberg, Elizabeth, 1976

Craig, Jenny, 1983
De Dominic, Patty, 1979
Demorest, Ellen Curtis, 1860
Dyson, Esther, 1983
Fertel, Ruth, 1965
Fields, Debbi, 1977
Graham, Bette Nesmith, 1954
Greiner, Helen, 2002
Gupta, Vinita, 1985
Handler, Ruth, 1959
Harper, Martha Matilda, 1891
Haughery, Margaret, 1840
Holdridge, Barbara, 1952
Hughes, Cathy L., 1999
Jenney, Mistress Sarah, 1644
Karan, Donna, 1984
Knox, Rose Markward, 1890
Lauder, Estée, 1946
Little, Carole, 1975
Lukens, Rebecca Webb, 1825
Malone, Annie, 1902
Mantell, Marianne, 1952
Matthias, Rebecca, 1982
McClintock, Jessica, 1969
Meskis, Joyce, 1974
Missett, Judi Sheppard, 1969
Musgrove, Mary, 1733
Natori, Josie Cruz, 1977
Nealey, Bertha, 1945
Nidetch, Jean, 1963
Ninham, Sandra, 1976
Philipse, Margaret Hardenbroeck, 1660
Pickford, Mary, 1916
Pinckney, Eliza Lucas, 1744
Pinkham, Lydia Estes, 1875
Poe, Sheri, 1987
Rosenthal, Ida, 1923
Rowland, Pleasant T., 1986
Rubinstein, Helena, 1914
Rudkin, Margaret, 1936
Siebert, Muriel, 1967
Sobrino, Maria de Lourdes, 1982
Stewart, Martha, 1997

Strong, Harriet, 1887
Tenderich, Gertrude, 1934
Vernon, Lillian, 1951
Vittadini, Adrienne, 1979
Walker, Madam C. J., 1905
Webster, Alma, 1976
Whitener, Catherine Evans, 1895
Winfrey, Oprah, 1985
Wise, Brownie, 1951
Wylie, Mary Washington, 1968

ENVIRONMENTALISTS
Carson, Rachel, 1962
Douglas, Marjory Stoneman, 1947
Patrick, Ruth, 1933
Sinclair, Mary, 1967

FASHION DESIGNERS AND INNOVATORS
Bloomer, Amelia, 1851
Bryant, Lena, 1904
Carnegie, Hattie, 1925
Claiborne Ortenberg, Elizabeth, 1976
Demorest, Ellen Curtis, 1860
Jacob, Mary Phelps, 1914
Karan, Donna, 1984
Keckley, Elizabeth, 1855
Little, Carole, 1975
Matthias, Rebecca, 1982
McClintock, Jessica, 1969
Natori, Josie Cruz, 1977
Vittadini, Adrienne, 1979

FILMMAKERS
Ephron, Nora, 1993
Foster, Jodie, 1991
Lupino, Ida, 1933
Marion, Frances, 1930
Marshall, Penny, 1988
Ono, Yoko, 1964
Parrish, Essie, 1941
Phillips, Julia, 1973
Streisand, Barbra, 1962
Taymor, Julie, 1997

FILM STARS. *See Actors*

FINANCIAL SERVICES PROVIDERS AND INVESTORS
Bay, Josephine Perfect, 1956
Cohen, Abby Joseph, 1998
Domini, Amy, 1989
Green, Hetty, 1916
Krawcheck, Sallie, 2004
Porter, Sylvia, 1942
Siebert, Muriel, 1967
Walker, Maggie Lena, 1903
Wylie, Mary Washington, 1968

FIRST LADIES
Adams, Abigail, 1776
Clinton, Hillary Rodham, 2000
Ford, Betty, 1974
Johnson, Lady Bird, 1965
Kennedy Onassis, Jacqueline, 1962
Madison, Dolley, 1844
Roosevelt, Eleanor, 1933

GOVERNORS
Grasso, Ella, 1975
Ray, Dixy Lee, 1976
Richards, Ann, 1988
Whitman, Christine Todd, 2004

HISTORIANS
Beard, Mary Ritter, 1946
Dawidowicz, Lucy, 1975
Faust, Drew Gilpin, 2007
Hamilton, Edith, 1930
Lerner, Gerda, 1981
Park, Maud Wood, 1901
Tuchman, Barbara, 1962
Warren, Mercy Otis, 1773

HORTICULTURISTS. *See Botanists and Horticulturists*

HUMANITARIANS
Addams, Jane, 1889

Barton, Clara, 1881
Bissell, Emily, 1907
Brinker, Nancy, 1982
Briones, Juana, 1844
Cole, Johnetta B., 1987
Cope, Mother Marianne, 1888
Dix, Dorothea, 1843
Eustis, Dorothy, 1929
Jacobs, Frances Wisebart, 1887
King, Coretta Scott, 1969
Neakok, Sadie, 1960
Shriver, Eunice Kennedy, 1961
Thomas, Marlo, 1966
Wauneka, Annie Dodge, 1963
Williams, Jody, 1997

INVENTORS
Abbott, Berenice, 1939
Adams, Maude, 1923
Ancker-Johnson, Betsy, 1966
Blanchard, Helen Augusta, 1873
Blodgett, Katherine, 1938
Chiang, Anne, 1984
Cochran, Josephine, 1886
Coston, Martha, 1871
Elion, Gertrude, 1988
Goode, Sarah, 1885
Graham, Bette Nesmith, 1954
Greene, Catherine, 1807
Greiner, Helen, 2002
Henry, Beulah, 1912
Jacob, Mary Phelps, 1914
Kies, Mary, 1809
Knight, Margaret, 1870
Kwolek, Stephanie, 1965
Masters, Sybilla, 1715
Montgomery, Mary Jane, 1864
Ochoa, Ellen, 1990
Pennington, Mary Engle, 1895
Pinckney, Eliza Lucas, 1744
Schroeder, Becky, 1974
Strong, Harriet, 1887
Whitener, Catherine Evans, 1895

Journalists

Bly, Nellie (Elizabeth Cochrane Seaman), 1889
Bonney, Thérèse, 1939
Bourke-White, Margaret, 1936
Cary, Mary Ann Shadd, 1853
Chappelle, Georgette (Dicky), 1943
Chung, Connie, 1983
Couric, Katie, 1990
Crier, Catherine, 1984
Croly, Jane Cunningham, 1889
Day, Dorothy, 1933
Ellerbee, Linda, 1972
Faludi, Susan, 1991
Goldman, Emma, 1889
Graham, Katharine Meyer, 1974
Harper, Ida Husted, 1922
Higgins, Maggie, 1951
House, Karen Elliot, 2002
Huxtable, Ada Louise, 1970
Idar, Jovita, 1911
Johnston, Frances Benjamin, 1889
Kilgallen, Dorothy, 1936
Laurie, Annie (Winifred Black), 1893
McBride, Mary Margaret, 1934
McDowell, Anna Elizabeth, 1855
McGrory, Mary, 1974
Pauley, Jane, 1976
Peña, Soledad, 1911
Porter, Sylvia, 1942
Roberts, Cokie, 1991
Sawyer, Diane, 1984
Stahl, Leslie, 1983
Stamberg, Susan, 1972
Thomas, Helen, 1974
Thompson, Dorothy, 1936
Walters, Barbara, 1976
Whitehead, Nancy Dickerson, 1960

Judges and Supreme Court Associate Justices

Allen, Florence, 1922
Cline, Genevieve, 1928
Crier, Catherine, 1984
Ginsburg, Ruth Bader, 1993
Motley, Constance Baker, 1943
O'Connor, Sandra Day, 1981

Labor Leaders

Bagley, Sarah, 1844
Beard, Mary Ritter, 1946
Chavez-Thompson, Linda, 1995
Flynn, Elizabeth Gurley, 1912
Huerta, Dolores, 1962
Jones, Mother (Mary Harris), 1903
Moskowitz, Belle, 1913
Nestor, Agnes, 1902
O'Sullivan, Mary Kenney, 1892
Parsons, Lucy (Lucia González), 1886
Roche, Josephine, 1912
Schneiderman, Rose, 1903
Tenayuca, Emma, 1938
Wright, Frances (Fanny), 1829

Lawyers

Abzug, Bella, 1971
Alexander, Sadie, 1935
Allen, Florence, 1922
Anderson, Violette, 1926
Bradwell, Myra, 1868
Brent, Margaret, 1648
Cary, Mary Ann Shadd, 1853
de Hoyos, Debora, 1991
Eastman, Crystal, 1910
Edelman, Marian Wright, 1973
Foltz, Clara Shortridge, 1878
Ginsburg, Ruth Bader, 1993
Jordan, Barbara, 1966
Lockwood, Belva, 1879
MacKinnon, Catharine Alice, 1979
Mansfield, Arabella, 1869
Motley, Constance Baker, 1943
Murray, Pauli, 1944
Norton, Eleanor Holmes, 1970
Norton, Gale, 2001
O'Connor, Sandra Day, 1981

Ray, Charlotte, 1872
Reno, Janet, 1993
Weddington, Sarah, 1973

Lecturers

Anthony, Susan B., 1869
Duniway, Abigail Scott, 1912
Fisher, Mary, 1992
Foster, Abby Kelley, 1838
Kanter, Rosabeth Moss, 1985
Keller, Helen, 1903
La Flesche Tibbles, Susette, 1880
Parsons, Lucy (Lucia González), 1886
Rand, Ayn, 1943
Roosevelt, Eleanor, 1933
Rose, Ernestine, 1840
Sanger, Margaret, 1916
Schaef, Anne Wilson, 1990
Shaw, Anna Howard, 1880
Stewart, Maria, 1831
Truth, Sojourner, 1843

Legislators

Abzug, Bella, 1971
Brathwaite-Burke, Yvonne, 1966
Chisholm, Shirley, 1968
Green, Edith, 1955
Jordan, Barbara, 1966
Norton, Eleanor Holmes, 1970
Pelosi, Nancy, 1987
Rankin, Jeannette, 1917
Rogers, Edith Nourse, 1925
Schroeder, Patricia, 1972
Smith, Margaret Chase, 1940
See also Senators

Mathematicians

Flugge-Lotz, Irmgard, 1960
Haynes, Euphemia Lofton, 1943
Hopper, Grace Murray, 1943
Ladd-Franklin, Christine, 1882
Noether, Amalie Emmy, 1921
Robinson, Julia, 1976

Medical Professionals. *See Dentists; Nurses; Physicians*

Military Personnel

Cobb, Jerrie, 1960
Cushman, Pauline, 1865
de Victor, Maude, 1978
Hallaren, Mary Agnes, 1948
Hobby, Oveta Culp, 1942
Holm, Jeanne M., 1971
Hopper, Grace Murray, 1943
Pitcher, Molly (Mary McCauley), 1778
Tompkins, Sally, 1861
Vaught, Wilma, 1966

Musicians

Beach, Amy Marcy Cheney, 1896
Carter, Maybelle, 1939
Carter, Regina, 2004
Dwyer, Doriot Anthony, 1952
Harrison, Hazel, 1915
Ono, Yoko, 1964
Powell, Maud, 1885
Ran, Shulamit, 1991
Urso, Camilla, 1852
Zwilich, Ellen, 1983
See also Composers; Conductors; Singers

Nobel Prize Winners

Addams, Jane, 1889
Balch, Emily Greene, 1946
Buck, Linda B., 2004
Buck, Pearl S., 1938
Cori, Gerty Radnitz, 1947
Elion, Gertrude, 1988
Goeppert-Mayer, Maria, 1963
McClintock, Barbara, 1983
Morrison, Toni, 1993
Williams, Jody, 1997
Yalow, Rosalyn, 1977

Nonprofit Organization Executives

Allen, Steffie, 1989

Bloch, Julia Chang, 1989
Bordas, Juana, 1977
Brinker, Nancy, 1982
Calderone, Mary Steichen, 1953
Dole, Elizabeth Hanford, 1983
Edelman, Marian Wright, 1973
Herrington, Alice, 1957
Jacobs, Frances Wisebart, 1887
Kuhn, Maggie, 1974
Lightner, Candy, 1980
Roulet, Sister Elaine, 1980
Sadker, Myra, 1994
Schwartz, Felice, 1962
Solomon, Hannah G., 1893
Wattleton, Faye, 1978
Wellington, Sheila, 1993

NURSES
Barton, Clara, 1884
Bickerdyke, Mary, 1862
Breckenridge, Mary, 1925
Church, Ellen, 1930
Livermore, Mary Rice, 1864
Tompkins, Sally, 1861
Wald, Lillian D., 1895

PAINTERS
Bachman, Maria Martin, 1833
Beaux, Cecilia, 1884
Cassatt, Mary, 1868
De Kooning, Elaine, 1963
Frankenthaler, Helen, 1950
Krasner, Lee, 1951
Moses, Grandma, 1939
O'Keeffe, Georgia, 1916
Peale, Anna Claypoole, 1824
Peale, Sarah Miriam, 1824
Perry, Lilla Cabot, 1889
Thomas, Alma, 1924

PHILANTHROPISTS
Cosby, Camille Olivia Hanks, 1987
Dodge, Grace Hoadley, 1886

Haughery, Margaret, 1840
Hunt, Swanee, 1997
Post, Marjorie Merriweather, 1929
Rubinstein, Helena, 1914
Smith, Sophia, 1871
Winfrey, Oprah, 1985

PHILOSOPHERS
Arendt, Hannah, 1951
Sontag, Susan, 1966

PHOTOGRAPHERS
Abbott, Berenice, 1939
Arbus, Diane, 1972
Bonney, Thérèse, 1939
Bourke-White, Margaret, 1936
Chappelle, Georgette (Dicky), 1943
Ettlinger, Marion, 1983
Johnston, Frances Benjamin, 1889
Lange, Dorothea, 1936
Leibovitz, Annie, 1970
Wolcott, Marion Post, 1938

PHYSICIANS
Alexander, Hattie, 1937
Apgar, Virginia, 1952
Blackwell, Elizabeth, 1849
Calderone, Mary Steichen, 1953
Caldicott, Helen, 1971
Cleveland, Emeline Horton, 1855
Cori, Gerty Radnitz, 1947
Crumpler, Rebecca Lee, 1864
Dick, Gladys, 1924
Elders, Jocelyn, 1993
Hamilton, Alice, 1947
Healy, Bernadine, 1991
Hunt, Harriot Kezia, 1835
Jacobi, Mary Putnam, 1872
Jemison, Mae, 1987
Kelsey, Frances, 1962
Love, Susan, 1990
Minoka-Hill, L. Rosa, 1899
Novello, Antonia, 1990
Pearce, Louise, 1949

Picotte, Susan La Flesche, 1889
Sabin, Florence, 1925
Seibert, Florence, 1941
Slye, Maud, 1911
Taussig, Helen Brooke, 1944
Walker, Mary Edwards, 1860
Zakrzewska, Marie Elizabeth, 1862

PHYSICISTS
Ancker-Johnson, Betsy, 1966
Blodgett, Katherine, 1938
Goeppert-Mayer, Maria, 1963
Jackson, Shirley Ann, 1973
Wu, Chien Shiung, 1944
Yalow, Rosalyn, 1977

PLAYWRIGHTS
Hansberry, Lorraine, 1959
Hellman, Lillian, 1952
Rowson, Susanna Haswell, 1790
Warren, Mercy Otis, 1773
Wasserstein, Wendy, 1989

POETS
Bates, Katherine Lee, 1895
Bradstreet, Anne, 1650
Brooks, Gwendolyn, 1949
Cisneros, Sandra, 1984
Crosby, Fanny, 1844
Dickinson, Emily, 1852
Dove, Rita, 1993
Howe, Julia Ward, 1862
Lazarus, Emma, 1883
Lorde, Audre (Gamba Adisa), 1991
Millay, Edna St. Vincent, 1923
Moore, Marianne, 1915
Prince, Lucy Terry, 1746
Rich, Adrienne, 1973
Rukeyser, Muriel, 1935
Warren, Mercy Otis, 1773
Wheatley, Phillis, 1773

POLITICIANS
Abzug, Bella, 1971

Brathwaite-Burke, Yvonne, 1966
Chisholm, Shirley, 1968
Clinton, Hillary Rodham, 2000
Ferraro, Geraldine, 1984
Grasso, Ella, 1975
Green, Edith, 1955
Jordan, Barbara, 1966
Kassebaum, Nancy Landon, 1978
Kelly, Sharon Pratt Dixon, 1991
Luce, Clare Boothe, 1953
Meir, Golda, 1969
Norton, Eleanor Holmes, 1970
Pelosi, Nancy, 1987
Rankin, Jeannette, 1917
Ray, Dixy Lee, 1976
Richards, Ann, 1988
Rogers, Edith Nourse, 1925
Schroeder, Patricia, 1972
Smith, Margaret Chase, 1940
Whitman, Christine Todd, 2001

POTTERS
Duckworth, Ruth, 1964
Martinez, Maria Montoya, 1919
Nampeyo, 1895
Wong, Jade Snow, 1950

PSYCHOLOGISTS
Follett, Mary Parker, 1896
Gilbreth, Lillian, 1931
Horney, Karen, 1939
Satir, Virginia, 1964
Westheimer, Ruth, 1970

PUBLIC SERVANTS
Abbott, Grace, 1921
Albright, Madeleine, 1997
Allen, Florence, 1922
Ancker-Johnson, Betsy, 1966
Anderson, Mary, 1920
Banuelos, Romana, 1949
Bronson, Ruth Muskrat, 1930
Clark, Georgia Neese, 1949
Cline, Genevieve, 1928

Davis, Alice, 1922
Douglas, Jennie, 1862
Earle, Sylvia Alice, 1990
Edelman, Marian Wright, 1973
Elders, Jocelyn, 1993
Field, Pattie, 1925
Ginsburg, Ruth Bader, 1993
Goddard, Mary Katherine, 1775
Harris, Patricia Roberts, 1979
Healy, Bernadine, 1991
Hobby, Oveta Culp, 1942
Kimbell, Gail, 2007
Kreps, Juanita, 1972
Lathrop, Julia, 1912
Lewis, Ida, 1879
Mankiller, Wilma, 1985
Motley, Constance Baker, 1943
Neakok, Sadie, 1960
Norton, Gale, 2001
Novello, Antonia, 1990
O'Connor, Sandra Say, 1981
Penn, Hannah Callowhill, 1746
Perkins, Frances, 1933
Peterson, Esther, 1961
Reno, Janet, 1993
Rice, Condoleezza, 1993
Roche, Josephine, 1912
Schulze, Tye Leung, 1910
Shalala, Donna, 1987
Temple Black, Shirley, 1935
Ward, Nancy, 1755
Webb, Wilma, 1998
Whitman, Christine Todd, 2001
Widnall, Sheila, 1993
See also Civic Leaders; Diplomats;
Politicians

PUBLIC SPEAKERS. *See Lecturers*

PUBLISHERS
Bradwell, Myra, 1868
Friedman, Jane, 2003
Goddard, Mary Katherine, 1775
Graham, Katharine Meyer, 1974

House, Karen Elliot, 2002
McDowell, Anna Elizabeth, 1855
Wallace, Lila, 1922
Woodhull, Victoria, 1872

RELIGIOUS LEADERS
Blackwell, Antoinette Brown, 1853
Cabrini, Mother Frances Xavier, 1889
Cope, Mother Marianne, 1888
Drexel, Sister Mary Katharine, 1891
Dyer, Mary, 1660
Eddy, Mary Baker, 1879
Hutchinson, Anne, 1637
Lee, Mother Ann, 1776
Murray, Pauli, 1944
Palmer, Phoebe Worrall, 1837
Parrish, Essie, 1941
Priesand, Sally J., 1972
Schori, Katharine Jefferts, 2006
Seton, Elizabeth Ann Bayley, 1797
Shaw, Anna Howard, 1880
Tekawitha, Kateri, 1676
Thering, Sister Rose, 1957
Towner, Margaret, 1956
White, Ellen, 1860

REPRESENTATIVES. *See Legislators*

REVOLUTIONARY WAR
  PERSONALITIES
Adams, Abigail, 1776
Corbin, Margaret, 1779
Fulton, Sarah, 1773
Goddard, Mary Katherine, 1775
Ludington, Sybil, 1777
Pitcher, Molly (Mary McCauley), 1778
Reed, Esther DeBerdt, 1780
Ross, Betsy, 1776
Sampson, Deborah, 1782
Warren, Mercy Otis, 1773

SCIENTISTS
Alexander, Hattie, 1937
Ancker-Johnson, Betsy, 1966

Bascom, Florence, 1893
Blodgett, Katherine, 1938
Brown, Rachel Fuller, 1950
Buck, Linda B., 2004
Burns, Stephanie, 2004
Cannon, Annie Jump, 1896
Carson, Rachel, 1962
Colden, Jane, 1757
Cori, Gerty Radnitz, 1947
Dick, Gladys, 1924
Earle, Sylvia Alice, 1990
Elion, Gertrude, 1988
Evans, Alice, 1917
Fleming, Williamina Stevens, 1906
Fossey, Dian, 1967
Furbish, Catherine, 1895
Goeppert-Mayer, Maria, 1963
Hazen, Elizabeth Lee, 1950
Hockfield, Susan, 2004
Jackson, Shirley Ann, 1973
Kwolek, Stephanie, 1965
Leavitt, Henrietta Swan, 1912
Logan, Martha Daniell, 1751
Maxwell, Martha, 1876
McClintock, Barbara, 1983
Mitchell, Maria, 1848
Patrick, Ruth, 1933
Pearce, Louise, 1919
Pennington, Mary Engle, 1895
Pool, Judith Graham, 1965
Ray, Dixy Lee, 1976
Richards, Ellen Swallow, 1882
Rubin, Vera, 1970
Sabin, Florence, 1925
Seibert, Florence, 1941
Semple, Ellen Churchill, 1924
Slye, Maud, 1911
Stevens, Nettie, 1905
Taussig, Helen Brooke, 1944
Theriot, Julie, 2004
Wong-Staal, Flossie, 1984
Wu, Chien Shiung, 1944
Yalow, Rosalyn, 1977

SCULPTORS
Duckworth, Ruth, 1964
Hesse, Eva, 1972
Hosmer, Harriet Goodhue, 1853
Hoxie, Vinnie Ream, 1866
Lewis, Edmonia, 1867
Lin, Maya, 1984
Nevelson, Louise, 1940
Taymor, Julie, 1997
Truitt, Anne, 1973
Wright, Patience, 1769

SENATORS
Clinton, Hillary Rodham, 2000
Dole, Elizabeth Hanford, 1983
Kassebaum, Nancy Landon, 1978
Smith, Margaret Chase, 1940

SINGERS
Anderson, Marian, 1939
Baez, Joan, 1960
Brice, Fanny, 1921
Carter, Maybelle, 1939
Cline, Patsy, 1957
Cruz, Celia, 1947
Estefan, Gloria, 1986
Fitzgerald, Ella, 1934
Franklin, Aretha, 1987
Garland, Judy, 1940
Holiday, Billie, 1935
Horne, Lena, 1943
Horne, Marilyn, 1969
Jackson, Mahalia, 1934
Jones, Sissieretta, 1888
King, Carole, 1971
Merman, Ethel, 1930
Midler, Bette, 1972
Parton, Dolly, 1975
Peters, Roberta, 1950
Price, Leontyne, 1952
Rainey, Ma, 1923
Reynolds, Malvina, 1950

Rivera, Chita, 1957
Sainte-Marie, Buffy, 1982
Shore, Dinah, 1955
Sills, Beverly, 1955
Smith, Bessie, 1919
Smith, Kate, 1926
Streisand, Barbra, 1962
Tucker, Sophie, 1906
Turner, Tina, 1956
Vaughan, Sarah, 1942
Waters, Ethel, 1927

SOCIAL ARBITERS
de Wolfe, Elsie, 1913
Post, Emily, 1922

SOCIAL REFORMERS
Abbott, Edith, 1908
Abbott, Grace, 1921
Addams, Jane, 1889
Atler, Marilyn Van Derbur, 1991
Blackwell, Alice Stone, 1890
Bordas, Juana, 1977
Breckenridge, Mary, 1925
Breckinridge, Sophonisba, 1901
Day, Dorothy, 1933
Dix, Dorothea, 1843
Eastman, Crystal, 1910
Eustis, Dorothy, 1929
Foster, Abby Kelley, 1838
Hunt, Harriot Kezia, 1835
Keller, Helen, 1903
Kelley, Florence, 1899
Lathrop, Julia, 1912
Moskowitz, Belle, 1913
Mott, Lucretia, 1840
Nation, Carry, 1899
Park, Maud Wood, 1901
Rose, Ernestine, 1840
Wald, Lillian D., 1895
Willard, Frances, 1874
*See also Civil Rights Activists; Humani-
tarians; Suffragists*

SOCIOLOGISTS
Balch, Emily Greene, 1946
Lee, Rose Hum, 1956

SPORTS EXECUTIVES
Alvarado, Linda, 1976
Bivens, Carolyn Vesper, 2005
Donahue, Margaret, 1950
Manley, Effa, 1935
Sarnoff, Ann, 2004
Stephens, Helen, 1936

SPORTS FIGURES. *See Athletes*

SUFFRAGISTS
Allen, Florence, 1922
Anthony, Susan B., 1869
Belmont, Alva Erskine Smith
    Vanderbilt, 1909
Blackwell, Alice Stone, 1890
Blair, Emily Newell, 1924
Blatch, Harriet Stanton, 1902
Bloomer, Amelia, 1851
Burns, Lucy, 1912
Catt, Carrie Chapman, 1900
Duniway, Abigail Scott, 1912
Foster, Abby Kelley, 1838
Gage, Matilda Joslyn, 1852
Howe, Julia Ward, 1862
Howland, Emily, 1857
Minor, Virginia, 1875
Mott, Lucretia, 1840
Nation, Carry, 1899
Park, Maud Wood, 1901
Paul, Alice, 1912
Shaw, Anna Howard, 1880
Stanton, Elizabeth Cady, 1848
Stone, Lucy, 1850
Wells-Barnett, Ida B., 1895
Woodward Pierce, Charlotte, 1920
Wright, Martha Coffin, 1833

SURGEON GENERALS
Elders, Jocelyn, 1993
Novello, Antonia, 1990

TEACHERS. *See Educators*

TELEVISION AND RADIO
PERSONALITIES
Allen, Gracie, 1911
Ball, Lucille, 1951
Berg, Gertrude, 1949
Burnett, Carol, 1959
Carroll, Diahann, 1968
Chen, Joyce, 1958
Child, Julia, 1963
Chung, Connie, 1983
Couric, Katie, 1990
Crier, Catherine, 1984
De Generes, Ellen, 1982
Ellerbee, Linda, 1972
Horwich, Frances, 1952
Kilgallen, Dorothy, 1936
McBride, Mary Margaret, 1934
Moore, Mary Tyler, 1970
O'Donnell, Rosie, 1986
Pauley, Jane, 1976
Rivers, Joan, 1975
Roberts, Cokie, 1991
Roberts, Robin, 1996
Sainte-Marie, Buffy, 1982
Sawyer, Diane, 1984
Shore, Dinah, 1955
Stahl, Leslie, 1983
Stamberg, Susan, 1972
Thomas, Marlo, 1966
Thompson, Dorothy, 1936
Tomlin, Lily, 1977
Walters, Barbara, 1976
Westheimer, Ruth, 1970
White, Betty, 1953
Whitehead, Nancy Dickerson, 1960
Winfrey, Oprah, 1985

TELEVISION EXECUTIVES,
PRODUCERS, AND WRITERS
Carsey, Marcy, 1981
Cooney, Joan Ganz, 1969

English, Diane, 1988
Hughes, Cathy L., 1999
Koplovitz, Kay, 1977
Laybourne, Geraldine, 1998
Mandabach, Caryn, 1986
McGrath, Judy, 2004
Winfrey, Oprah, 1985

VISUAL ARTISTS
Bachman, Maria Martin, 1833
Beaux, Cecilia, 1884
Cassatt, Mary, 1868
Chicago, Judy, 1979
De Kooning, Elaine, 1963
Duckworth, Ruth, 1964
Frankenthaler, Helen, 1950
Goodacre, Glenna, 1993
Hesse, Eva, 1972
Hosmer, Harriet Goodhue, 1853
Hoxie, Vinnie Ream, 1866
Krasner, Lee, 1951
Lewis, Edmonia, 1867
Lin, Maya, 1981
Messick, Dale, 1940
Moses, Grandma, 1939
Nevelson, Louise, 1940
O'Keeffe, Georgia, 1916
Ono, Yoko, 1964
Palmer, Frances (Fanny) Bond, 1849
Peale, Anna Claypoole, 1824
Peale, Sarah Miriam, 1824
Perry, Lilla Cabot, 1889
Qoyawayma, Polingaysi, 1925
Taymor, Julie, 1997
Terry, Hilda, 1941
Thomas, Alma, 1924
Truitt, Anne, 1973
Wright, Patience, 1769
See also Photographers; Potters

WOMEN'S RIGHTS ACTIVISTS
Anthony, Susan B., 1869
Blackwell, Alice Stone, 1890

Catt, Carrie Chapman, 1900
Chicago, Judy, 1979
Duniway, Abigail Scott, 1912
Friedan, Betty, 1963
Gage, Matilda Joslyn, 1852
Grimké, Angelina, 1838
Minor, Virginia, 1875
Mott, Lucretia, 1840
Pogrebin, Letty Cottin, 1971
Rose, Ernestine, 1840
Shaw, Anna Howard, 1880
Stanton, Elizabeth Cady, 1848
Steinem, Gloria, 1971
Stone, Lucy, 1850
Woodhull, Victoria, 1872
Woodward Pierce, Charlotte, 1920
Wright, Frances (Fanny), 1829
Wright, Martha Coffin, 1833
See also Suffragists

WRITERS

Abbott, Berenice, 1939
Abbott, Edith, 1908
Alcott, Louisa May, 1868
Angelou, Maya, 1972
Anthony, Susan B., 1869
Antin, Mary, 1912
Anzaldúa, Gloria, 1987
Arbuthnot, May, 1940
Blair, Emily Newell, 1924
Bombeck, Erma, 1968
Brown, Margaret Wise, 1937
Brownmiller, Susan, 1975
Buck, Pearl S., 1938
Butler, Octavia, 1979
Cather, Willa, 1918
Cheng, Nien, 1986
Chesnut, Mary, 1905
Child, Lydia Maria, 1833
Chopin, Kate, 1889
Cisneros, Sandra, 1984
Converse, Harriet Maxwell, 1891
Cooper, Anna Julia, 1892

de Wolfe, Elsie, 1913
Didion, Joan, 1968
Dodge, Mary Mapes, 1865
Dove, Rita, 1993
Dworkin, Andrea, 1974
English, Diane, 1988
Ephron, Nora, 1993
Faludi, Susan, 1991
Farnham, Eliza Wood Burhans, 1864
Farrar, Margaret Petherbridge, 1924
Ferber, Edna, 1924
Friedan, Betty, 1963
Fuller, Margaret, 1839
Gilman, Charlotte Perkins, 1892
Glasgow, Ellen, 1925
Hale, Sarah Josepha, 1837
Hamilton, Edith, 1930
Hamilton, Virginia, 1975
Hansberry, Lorraine, 1959
Harper, Frances E. W., 1860
Harper, Ida Husted, 1922
Helgesen, Sally, 1990
Hellman, Lillian, 1952
Howe, Julia Ward, 1862
Hurston, Zora Neale, 1939
Jacobs, Jane, 1961
Jewett, Sarah Orne, 1896
Kanter, Rosabeth Moss, 1985
Keller, Helen, 1903
Kingston, Maxine Hong, 1976
Knight, Sarah Kemble, 1704
Kübler-Ross, Elisabeth, 1969
Lee, Harper, 1960
Lee, Rose Hum, 1956
Lerner, Gerda, 1981
Lindbergh, Anne Morrow, 1930
Loos, Anita, 1912
Lord, Bette Bao, 1981
Lorde, Audre (Gamba Adisa), 1991
Love, Susan, 1990
MacLaine, Shirley, 1955
Millay, Edna St. Vincent, 1923
Millett, Kate, 1970

Mitchell, Margaret, 1936
Moore, Marianne, 1915
Morrison, Toni, 1993
Mourning Dove, 1927
Murray, Judith Sargent Stevens, 1790
Oates, Joyce Carol, 1970
O'Connor, Flannery, 1952
Olsen, Tillie, 1961
Parker, Dorothy, 1927
Porter, Katherine Anne, 1962
Rand, Ayn, 1943
Rawlings, Marjorie Kinnan, 1938
Rich, Adrienne, 1973
Roberts, Cokie, 1991
Roosevelt, Eleanor, 1933
Rowlandson, Mary, 1682
Rowson, Susanna Haswell, 1790
Sadker, Myra, 1994
Sandoz, Mari, 1935
Saubel, Katherine Siva, 1987
Sawyer, Ruth, 1915
Schaef, Anne Wilson, 1990
Schwartz, Felice, 1962
Sheehy, Gail, 1992
Silko, Leslie, 1977
Smith, Margaret Bayard, 1824
Sontag, Susan, 1966
Stein, Gertrude, 1909
Stowe, Harriet Beecher, 1852
Tan, Amy, 1989
Tarbell, Ida, 1904
Uchida, Yoshiko, 1949
Walker, Alice, 1982
Warren, Mercy Otis, 1773
Wellington, Sheila, 1993
Wells-Barnett, Ida B., 1895
Welty, Eudora, 1972
Wharton, Edith, 1920
Wheatley, Phillis, 1773
Wilder, Laura Ingalls, 1935
Wilson, Harriet, 1859
Wong, Jade Snow, 1950
Yamamoto, Hisaye, 1950

Zitkala Sa (Red Bird), 1901
See also Editors; Journalists; Poets

YOUTH GROUP FOUNDERS

Field Shambaugh, Jessie, 1904
Gulick, Charlotte Vetter, 1910
Low, Juliette Gordon, 1912

Acknowledgments

Every book is clearly not only the product of its authors. Ours is no different. Many, many women, and a number of men, helped us immeasurably. Our deepest thanks go first and foremost to Jill Marce, who brought us together. Early supporters include many women from the Women's Vision Foundation. They are to be thanked for their unwavering support throughout this book project, especially Steffie Allen, Linda Bedinger, Christie Doherty, and Joan Miller.

To the women of CH2M Hill, and especially Joan Miller again, we offer our gratitude. Thanks to Antonette De Lauro, Nicole Trousdale, Kathy Ehler, Judy Alvis, Melissa Linn, Melanie Moore, and Lindsay Paiz; their creativity gave form to the timeline's early pictorial display.

We offer our appreciation to former secretary of state Madeleine Albright, whose support for women was extended to us as authors; thanks as well to her associate at the Albright Group, Jamie Smith.

Thanks to Patricia McLaughlin, whose PR helped launch our timeline project; to Lucy Strupp from Milestone Marketing, Inc., who assisted with our traveling exhibits; to John Gallagher and his colleagues at Rich Frog, who produced our customized ruler; to Karyn Ruth White, who recommended us to a book agent; to Robert W. Taylor, for his interest, creativity, and initial book design; and to Carolyn Miller, who helped us initially with the gargantuan task of image permissions.

We are reminded of the many women over the years who have told us their stories, alerted us to important women for inclusion, showed us their support, and given us their appreciation; we sincerely thank each of you.

The process brought us to our book agent, Sandra Bond, who never stopped believing; we are most appreciative. Our deepest thanks are to the incredible HarperCollins team–Phil Friedman, Lisa Hacken, and Laura Klynstra. Our gratitude goes to Lisa, who tightened and focused our material. Laura, especially, must be singled out for her talent, her magnificent visual vision of our material, and her creative drive and overall attention to every detail of this project; each of her drafts and each of her ideas brought amazement from us both. We truly appreciate the efforts of Maureen O'Neal, whose editorial prowess gave new direction to our book.

Lastly, a huge thank-you goes to our husbands, Jim Ottinger and David Tietjen, for their patience, love, and support.